Fundamentals of Geophysics

Fundamentals of Geophysics

Summer Murphy

MURPHY & MOORE
www.murphy-moorepublishing.com

Fundamentals of Geophysics
Summer Murphy
ISBN: 978-1-63987-246-6 (Hardback)

MURPHY & MOORE

Published by Murphy & Moore Publishing,
1 Rockefeller Plaza,
New York City, NY 10020, USA

Cataloging-in-Publication Data

Fundamentals of geophysics / Summer Murphy.
p. cm.
Includes bibliographical references and index.
ISBN 978-1-63987-246-6
1. Geophysics. 2. Earth sciences. 3. Physics. I. Murphy, Summer.
QE501 .G46 2022
550--dc23

For more information regarding Murphy & Moore Publishing and its products, please visit the publisher's website www.murphy-moorepublishing.com

TABLE OF CONTENTS

Preface VII

Chapter 1 Introduction 1

- Geophysics 1
- Geodesy 6
- Gravity 7
- Gravimetry 27
- Earth's Radioactivity 37

Chapter 2 Sub-fields of Geophysics 42

- Environmental Geophysics 42
- Exploration Geophysics 63
- Borehole Geophysics 67
- Biogeophysics 73
- Near-surface Geophysics 101
- Tectonophysics 104
- Hydrogeophysics 105

Chapter 3 Geodynamics 118

- Geodynamics of Terrestrial Exoplanets 130
- Geothermal Gradient 137
- Mineral Physics 142
- Dynamic Topography 146

Chapter 4 Geomagnetism and Paleomagnetism 151

- Geomagnetism 151
- Magnetic Declination 171
- Magnetic Inclination 180
- Dipole Model of the Earth's Magnetic Field 181
- The Dynamo Effect 183

- Geomagnetic Reversal 191
- Magnetic Anomaly 198
- Paleomagnetism 206
- Principles of Remanent Magnetization 208
- Paleomagnetic Poles 211

Chapter 5 Plate Tectonics 218

- Tectonic Plate 218
- Plate Tectonics 222
- Plate Boundaries 226
- Plate Reconstruction 228

Permissions

Index

PREFACE

It is with great pleasure that I present this book. It has been carefully written after numerous discussions with my peers and other practitioners of the field. I would like to take this opportunity to thank my family and friends who have been extremely supporting at every step in my life.

The branch of natural science which deals with the physical processes and properties of Earth and its surrounding environment is known as geophysics. It focuses on the study of gravitational and magnetic fields, Earth's shape and its internal structure, dynamics of plate tectonics and rock formation. It is also used to meet societal needs such as environmental protection, mineral resources and mitigation of natural hazards. This field uses concepts from physics to study the internal structure of the earth. Gravity, electromagnetic waves, radioactivity, mineral physics and heat flow are some of the areas of study within this discipline. It uses a wide variety of methods such as electrical method, gravity and magnetic surveys, and seismic and magneto-telluric surveys. Most of the topics introduced in this book cover new techniques and the applications of this field. It presents the complex subject of geophysics in the most comprehensible and easy to understand language. This book is an essential guide for both academicians and those who wish to pursue this discipline further.

The chapters below are organized to facilitate a comprehensive understanding of the subject:

Chapter - Introduction

Geophysics is the field of natural science that deals with the physical processes and properties of the Earth and its surrounding environment. It also includes water cycle, fluid dynamics of the oceans and the atmosphere and study of moon and other planets. This is an introductory chapter which will briefly introduce about geophysics.

Chapter - Sub-fields of Geophysics

Geophysics is branched into a number of sub-fields such as exploration geophysics, geodynamics, tectonophysics, borehole geophysics, hydrogeophysics, environmental geophysics, etc. This chapter closely examines these sub-fields of geophysics to provide an extensive understanding of the subject.

Chapter - Geodynamics

The sub-field of geophysics that deals with the dynamics and geologic phenomena of the Earth such as seafloor spreading, mountain building, volcanoes, earthquakes and faulting, is termed as geodynamics. This chapter has been carefully written to provide an extensive understanding of the geodynamics.

Chapter - Geomagnetism and Paleomagnetism

The dynamics of the Earth's magnetic field are studied under geomagnetism. Paleomagnetism is associated with the magnetism in rocks induced by the Earth's magnetic field at the time of formation. The topics elaborated in this chapter will help in gaining a better perspective of the concepts related to geomagnetism and paleomagnetism.

Chapter - Plate Tectonics

Plate tectonics is defined as a scientific theory which describes the large scale motion and movements of the plates of the Earth's lithosphere. A number of forces drive the movement of tectonic plates. Three of the major driving forces of tectonic plates are mantle convection currents, ridge push and slab pull. This chapter discusses the subject of plate tectonics in detail.

Summer Murphy

Introduction 1

- **Geophysics**
- **Geodesy**
- **Gravity**
- **Gravimetry**
- **Earth's Radioactivity**

Geophysics is the field of natural science that deals with the physical processes and properties of the Earth and its surrounding environment. It also includes water cycle, fluid dynamics of the oceans and the atmosphere and study of moon and other planets. This is an introductory chapter which will briefly introduce about geophysics.

Geophysics

Geophysics is a science that derives from geology and deals with the study of the physical properties of the Earth. It Includes aspects such as the research of the internal composition of the planet, the heat flow from the inside of the Earth, the force of gravity that forms the gravitational field, the attraction magnetic force exerted by an ideal magnet inside the Earth creates the geomagnetic field, and propagation of seismic waves through the rocks of earth's crust.

It helps to an appropriate location for civil works and the prevention of natural disasters; it also optimizes the process of exploration and extraction of minerals, water and energy. Through the register of different types of seismic waves (indirect method of geophysics), has been possible to define the internal composition of the Earth, dividing it in layers with thickness and contents specifications, without necessarily being in sight of human being.

Nowadays, the most accepted theory establishes that when the Earth began its evolution –about 4,700 million years ago, its initial temperature was 1,000° C; that temperature began to increase inward due to energy derived from radioactive minerals that as a result of their weight sank toward the center of the planet. The surface then started to cool with the help of external meteoric agents.

By increasing the internal temperature and reaching the melting point of iron and other minerals is how it begins the planet differentiation into layers of different density and therefore other mineral content, staying in the center of the planet most of the heavy minerals and to the outside most of the lighter minerals; thus is accomplished to differentiate layers of the planet in the inner core, outer core, lower mantle, asthenosphere or upper mantle, lithosphere and within this, a thin outer shell called earth crust inhabited by living beings that constitute the biosphere.

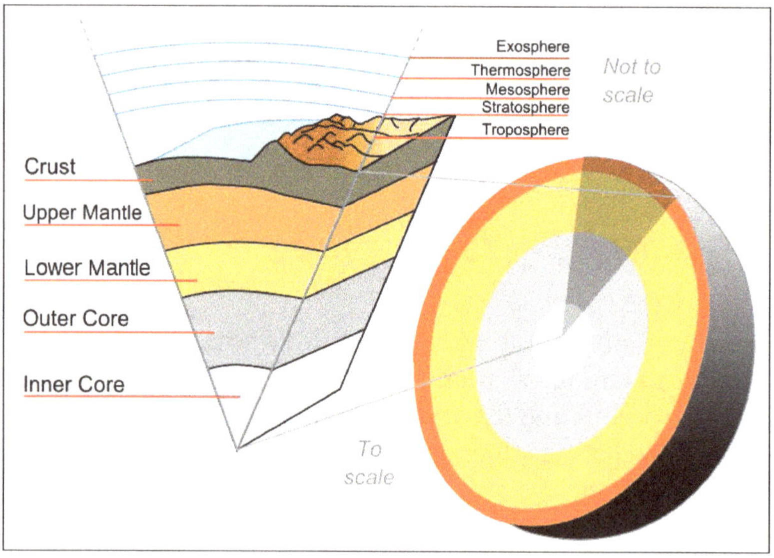

Heat flux is energy in transit and flows from high temperature regions inside the Earth towards low temperature regions located on the surface of the planet, creating convection currents in the molten rock (magma), which in some cases are manifested by creating volcanic areas. When they occur in aperture areas in the sea, oceanic ridges or middle oceanic are formed.

Geothermal Energy and its Application

Geophysics rests in other branches such as geothermal: science related to the study of the internal heat from the Earth and it is used in the location of natural hot water reservoirs.

Geothermal energy is manifested as a vibration of atoms whose intensity determines the temperature and reaches the surface by diffusion, by convective movements and by water circulation in the deep. This energy has its origin in the initial temperature of the planet; the second source is radioactivity derived from elements such as thorium, uranium and potassium - which are the minerals in rocks-, and as a third source, the gravity force into thermal energy during the compression process and the planet differentiation.

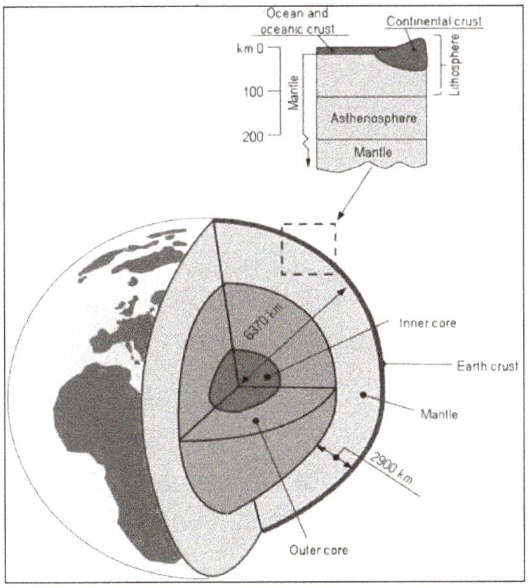

Its use is intended for the generation of electricity as steam produced by the natural hot geothermal fluids, is an alternative to that obtained in power plants by burning fossil material or by nuclear fission.

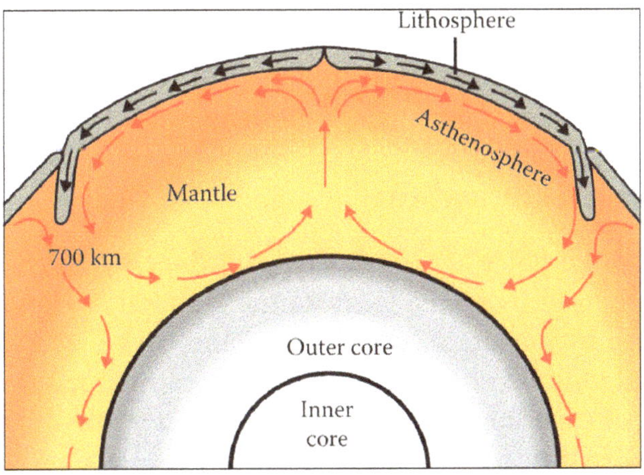

Heat flow in the interior of the earth.

Modern drilling in geothermal systems reaches water reserves and steam heated by deep magmas, which are up to 3,000 meters below sea level. The vapor is purified in the wellhead before being transported in large pipes and isolated to the turbines. Thermal energy can also be obtained from any cleft and geyser.

Gravitational and Magnetic Fields

Earth is influenced by two fields of internal origin, affecting all the processes occurring on the surface of the planet, these are:

- Gravitational field: Is that in which any body attracts others through their gravitational interaction fields, being possible to calculate the gravity effects such as: the acceleration of a falling object, the orbits of planets, the trajectory of a thrown object or an aerial vehicle.

The gravity local value (G) corresponds to the period –of a full swing time – of a rotating pendulum or, the acceleration of a falling weight.

- The magnetic field or geomagnetic: Is based on the magnetic properties of iron ore called magnetite, also known as "loaded stone" from which the Chinese invented the first magnetic compass that consisted on a rope suspended from a piece of this mineral.

  This field is described as a model of a small but powerful permanent magnetic bar, located near the center of the Earth and inclined about 11° from the geographic axis; the force lines of the magnetic field indicate the presence of a magnetic force at every point in space.

  The field has intensity and direction; its strength depends on the amount of magnetic material, the distance and direction to the detector. A magnetized needle, free to rotate on a pivot in a horizontal plane will rotate to a parallel position to the local power line, approximately north-south direction, denominating the end pointing to the geographic North, magnetic north pole and will be slightly off to the East or to the West of the geographic North Pole, depending on the location of the observer. This deviation angle of the geographic North is called declination.

Measurement and Exploration

Exploration of "potential fields" refers to the work done by magnetic and gravimetric procedures; both are used by specialists to understand and to interpret the behavior of the earth's surface by magnetic susceptibility and rock density. These geophysical methods correspond to indirect tools that help to locate structural features, as faults, fractures, certain rock types and geologic basement. They are also useful in the search for mineral deposits, hydrocarbons and aquifers.

Geophysicists' measure magnetic intensity with a tool called a magnetometer, accurate to 0.1 nT. The measure unit of magnetic field strength is nanotesla (nT), and differs according to the region of the globe, for example, ranges from 25,000 nT in the equator to 70,000 nT at the poles. A nanotesla equals 1 gamma, and 105 gammas equals to 1 oersted.

The magnetometer measures all the effects of earth's magnetic field and slow changes are updated every five years by the International Geomagnetic Reference Field (IGRF), so that data is associated to the organism.

First magnetometry studies are performed with aerial scan (aeromagnetometry), later in specific areas terrestrial magnetic scanning is performed, focused on accurate localization of ferriferous mineral deposits or any metallic deposit, containing magnetic minerals (magnetite, pyrrhotite).

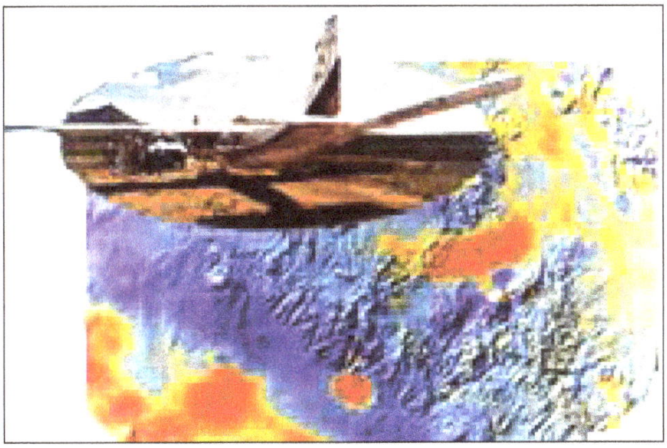

Geophysics.

Magnetic Field Variations

To study the magnetic properties of the rocks, geophysics is based on the discipline called paleomagnetism, also used to show the movement of masses of rock through geologic time. For example, this type of analysis has shown that the North and South Poles have been reversed over the geological years.

Through geologic time, both direction and intensity have been observed with variations in the magnetic field of the Earth. Just like all dipole magnets, the Earth has a magnetic field (core or main field) corresponding to the North and South Poles, therefore the angle between a compass and the true North is called the magnetic declination.

The magnetic field is affected by short variations due to magnetic storms. Even when the strength of earth's magnetic field is not so high, it is able to magnetize some kinds of rocks that contain iron or some other magnetic levels.

The difference between the measured magnetic values and the model of earth core values are known as "magnetic anomalies", mainly caused by variations in the magnetization of crustal rocks and due to the dipolar nature of the phenomenon, they can be positive or negative.

Most sedimentary rocks are nonmagnetic, igneous rocks rich in iron ores have high magnetic susceptibility or high magnetism (property that describes the amount of magnetic material in a lithological unit).

Rocks have magnetic minerals with two kinds of magnetization:

- Induced: Refers to the presence of an external magnetic field.
- Remnant: This is trapped inside the rock.

Geodesy

Geodesy is the science which deals with the methods of precise measurements of elements of the surface of the earth and their treatment for the determination of geographic positions on the surface of the earth. It also deals with the theory of size and shape of the earth. Geodesy may be broadly divided into two branches, namely:

- Geometric Geodesy.
- Physical Geodesy.

There is third one also, known as Satellite Geodesy. Thus geometric geodesy appears to be purely geometrical science as it deals with the geometry (shape and size) of the earth.

Determination of geographical positions on the surface of earth can be made by observing celestial bodies, and thus comes under geodetic astronomy, but this can be included under geometric geodesy.

Earth gravity field is a physical entity and is involved in most of the geodetic measurements, even the purely geometric ones. The measurement of geodetic astronomy, triangulation and leveling, all make essential use of plumb line being the direction of gravity vector.

Thus, astro-geodetic methods which use astro determination of latitude, longitude, and azimuth and geodetic operations e.g. triangulation, trilateration, base measurement etc., may be considered as belonging to Physical geodesy fully as much as the gravimetric methods.

As a general distinction astro-geodetic methods come under geometric geodesy which use the direction of gravity vector, employing geometrical techniques, where as the gravimetric methods come under physical geodesy, which operate with the magnitude of 'g' using potential theory. A sharp demarcation is impossible and there are frequent overlaps.

Satellite Geodesy: Satellite geodesy comprises the observational and computational techniques which allow the solution of geodetic problems by the use of precise measurements to, from or between artificial , mostly near the earth satellites.

Applications of Geodesy

- Primary or zero order triangulation, trilateration and traverse.
- The measurement of height above sea-level by triangulation or sprit levelling.

- Astronomically observations of latitude, longitude and azimuth to locate origins of surveys, and to control their direction.
- Crustal movements. To detect changes in the relative positions on the ground, and in their heights above sea level.
- Observation of the direction of Gravity by astronomical observations for latitude and longitude.
- Observation of the intensity of Gravity by the pendulum and other apparatus.
- To deduce the exact form of earth's sea level equipotential surfaces at all heights.
- Polar motion studies.
- Earth tides.
- The separation between the Geoid and the mean sea level.
- Engineering surveys.
- Satellite geodesy: includes the modern techniques of positioning by space methods e.g. GPS, SLR and VLBI etc.

Gravity

Gravity is also called as gravitation in mechanics is the universal force of attraction acting between all matter. It is by far the weakest known force in nature and thus plays no role in determining the internal properties of everyday matter. On the other hand, through its long reach and universal action, it controls the trajectories of bodies in the solar system and elsewhere in the universe and the structures and evolution of stars, galaxies, and the whole cosmos. On Earth all bodies have a weight, or downward force of gravity, proportional to their mass, which Earth's mass exerts on them. Gravity is measured by the acceleration that it gives to freely falling objects. At Earth's surface the acceleration of gravity is about 9.8 metres (32 feet) per second per second. Thus, for every second an object is in free fall, its speed increases by about 9.8 metres per second. At the surface of the Moon the acceleration of a freely falling body is about 1.6 metres per second per second.

In this picture a galactic cluster, about five billion light-years away, produces a tremendous gravitational field that "bends" light around it. This lens produces multiple copies of a blue galaxy about twice as distant. Four images are visible in a circle surrounding the lens; a fifth is visible near the centre of the picture.

The works of Isaac Newton and Albert Einstein dominate the development of gravitational theory. Newton's classical theory of gravitational force held sway from his Principia, published in 1687, until Einstein's work in the early 20th century. Newton's theory is sufficient even today for all but the most precise applications. Einstein's theory of general relativity predicts only minute quantitative differences from the Newtonian theory except in a few special cases. The major significance of Einstein's theory is its radical conceptual departure from classical theory and its implications for further growth in physical thought.

Gravitational lens, as observed by the Hubble Space Telescope.

The launch of space vehicles and developments of research from them have led to great improvements in measurements of gravity around Earth, other planets, and the Moon and in experiments on the nature of gravitation.

Newton's Law of Gravity

Newton discovered the relationship between the motion of the Moon and the motion of a body falling freely on Earth. By his dynamical and gravitational theories, he explained Kepler's laws and established the modern quantitative science of gravitation. Newton assumed the existence of an attractive force between all massive bodies, one that does not require bodily contact and that acts at a distance. By invoking his law of inertia (bodies not acted upon by a force move at constant speed in a straight line), Newton concluded that a force exerted by Earth on the Moon is needed to keep it in a circular motion about Earth rather than moving in a straight line. He realized that this force could be, at long range, the same as the force with which Earth pulls objects on its surface downward. When Newton discovered that the acceleration of the Moon is 1/3,600 smaller than the acceleration at the surface of Earth, he related the number 3,600 to the square of the radius of Earth. He calculated that the circular orbital motion of radius R and period T requires a constant inward acceleration A equal to the product of $4\pi^2$ and the ratio of the radius to the square of the time:

$$A = \frac{4\pi^2 R}{T^2}$$

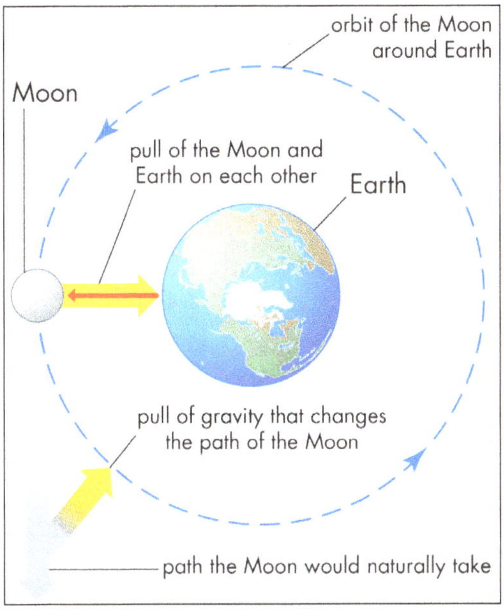

Effects of gravity on Earth and the Moon.

The Moon's orbit has a radius of about 384,000 km (239,000 miles; approximately 60 Earth radii), and its period is 27.3 days (its synodic period, or period measured in terms of lunar phases, is about 29.5 days). Newton found the Moon's inward acceleration in its orbit to be 0.0027 metre per second per second, the same as $(1/60)^2$ of the acceleration of a falling object at the surface of Earth.

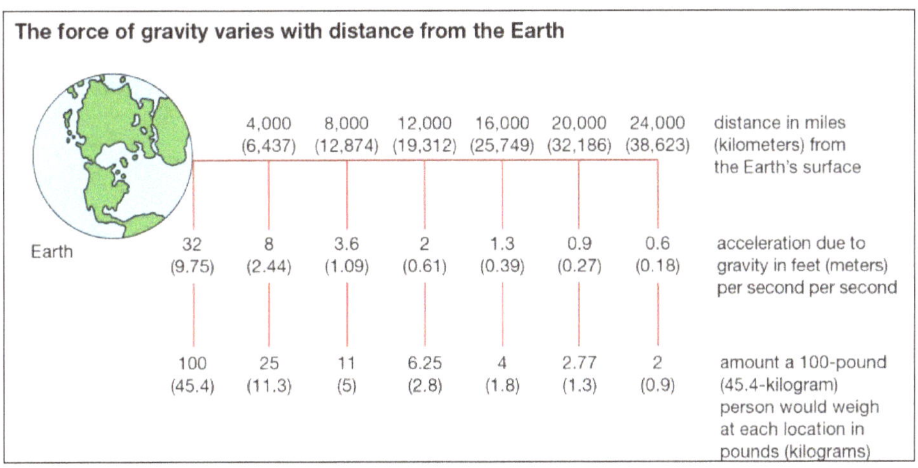

Gravitational force.

In Newton's theory every least particle of matter attracts every other particle gravitationally, and on that basis he showed that the attraction of a finite body with spherical symmetry is the same as that of the whole mass at the centre of the body. More generally, the attraction of any body at a sufficiently great distance is equal to that of the whole mass at the centre of mass. He could thus relate the two accelerations, that of the Moon and that of a body falling freely on Earth, to a common interaction, a gravitational force between bodies that diminishes as the inverse square of the distance between them. Thus, if the distance between the bodies is doubled, the force on them is reduced to a fourth of the original.

Newton saw that the gravitational force between bodies must depend on the masses of the bodies. Since a body of mass M experiencing a force F accelerates at a rate F/M, a force of gravity proportional to M would be consistent with Galileo's observation that all bodies accelerate under gravity toward Earth at the same rate, a fact that Newton also tested experimentally. In Newton's equation:

$$F_{12} = \frac{GM_1M_2}{r_{12}^2}$$

F_{12} is the magnitude of the gravitational force acting between masses M_1 and M_2 separated by distance r_{12}. The force equals the product of these masses and of G, a universal constant, divided by the square of the distance.

The constant G is a quantity with the physical dimensions $(\text{length})^3/(\text{mass})(\text{time})^2$; its numerical value depends on the physical units of length, mass, and time used.

The force acts in the direction of the line joining the two bodies and so is represented naturally as a vector, F. If r is the vector separation of the bodies, then:

$$\mathbf{F} = \frac{GM_1M_2\mathbf{r}}{r^3}$$

In this expression the factor r/r^3 acts in the direction of r and is numerically equal to $1/r^2$.

The attractive force of a number of bodies of masses M_1 on a body of mass M is:

$$\mathbf{F} = \frac{GM\sum_1 M_1\mathbf{r}_1}{r_1^3}$$

where Σ_1 means that the forces because of all the attracting bodies must be added together vectorially. This is Newton's gravitational law essentially in its original form. A simpler expression, equation below, gives the surface acceleration on Earth. Setting a mass equal to Earth's mass M_E and the distance equal to Earth's radius r_E, the downward acceleration of a body at the surface g is equal to the product of the universal gravitational constant and the mass of Earth divided by the square of the radius:

$$g = \frac{GM_E}{r_E^2}.$$

Weight and Mass

The weight W of a body can be measured by the equal and opposite force necessary to prevent the downward acceleration; that is M_g. The same body placed on the surface of the Moon has the same mass, but, as the Moon has a mass of about $^1/_{81}$ times that of Earth and a radius of just 0.27 that of Earth, the body on the lunar surface has a weight of only $^1/_6$ its Earth weight, as the Apollo program astronauts demonstrated. Passengers and instruments in orbiting satellites are in free fall. They experience weightless conditions even though their masses remain the same as on Earth.

Equations $A = \frac{4\pi^2 R}{T^2}$ and $F_{12} = \frac{GM_1M_2}{r_{12}^{\ 2}}$ can be used to derive Kepler's third law for the case of circular planetary orbits. By using the expression for the acceleration A in equation $A = \frac{4\pi^2 R}{T^2}$ for the force of gravity for the planet GM_PM_S/R^2 divided by the planet's mass MP, the following equation, in which M_S is the mass of the Sun, is obtained:

$$\frac{GM_S}{R^2} = \frac{4\pi^2 R}{T^2} \quad \text{or} \quad R^3 = \left(\frac{GM_S}{4\pi^2}\right)T^2$$

Kepler's very important second law depends only on the fact that the force between two bodies is along the line joining them.

Newton was thus able to show that all three of Kepler's observationally derived laws follow mathematically from the assumption of his own laws of motion and gravity. In all observations of the motion of a celestial body, only the product of G and the mass can be found. Newton first estimated the magnitude of G by assuming Earth's average mass density to be about 5.5 times that of water (somewhat greater than Earth's surface rock density) and by calculating Earth's mass from this. Then, taking M_E and r_E as Earth's mass and radius, respectively, the value of G was:

$$G = \frac{g r_E^{\ 2}}{M_E}$$

which numerically comes close to the accepted value of $6.6726 \times 10^{-11}\ m^3\ s^{-2}\ kg^{-1}$, first directly measured by Henry Cavendish.

Comparing equation $g = \frac{GM_E}{r_E^2}$ for Earth's surface acceleration g with the R^3/T^2 ratio for the planets, a formula for the ratio of the Sun's mass M_S to Earth's mass M_E was obtained in terms of known quantities, R_E being the radius of Earth's orbit:

$$\frac{M_S}{M_E} = \frac{4\pi^2 R_E^3}{g T_E^2 r_E^2} \cong 325{,}000$$

The motions of the moons of Jupiter around Jupiter obey Kepler's laws just as the planets do around the Sun. Thus, Newton calculated that Jupiter, with a radius 11 times larger than Earth's, was 318 times more massive than Earth but only $^1/_4$ as dense.

Interaction between Celestial Bodies

When two celestial bodies of comparable mass interact gravitationally, both orbit about a fixed point (the centre of mass of the two bodies). This point lies between the bodies on the line joining them at a position such that the products of the distance to each body with the mass of each body are equal. Thus, Earth and the Moon move in complementary orbits about their common centre of mass. The motion of Earth has two observable consequences. First, the direction of the Sun as seen from Earth relative to the very distant stars varies each month by about 12 arc seconds in addition to the Sun's annual motion. Second, the line-of-sight velocity from Earth to a freely moving spacecraft varies each month by 2.04 metres per second, according to very accurate data obtained from radio tracking. From these results the Moon is found to have a mass $^1/_{81}$ times that of Earth. With slight modifications Kepler's laws remain valid for systems of two comparable masses; the foci of the elliptical orbits are the two-body centre-of-mass positions, and, putting $M_1 + M_2$ instead of M_S in the expression of Kepler's third law, equation $R^3 = \left(\frac{GM_S}{4\ ^{2}}\right)T^2$, the third law reads:

$$R^3 = \frac{G(M_1 + M_2)}{4\pi^2}T^2$$

That agrees with equation $R^3 = \left(\frac{GM_S}{4\pi^2}\right)T^2$ when one body is so small that its mass can be neglected. The rescaled formula can be used to determine the separate masses of binary stars (pairs of stars orbiting around each other) that are a known distance from the solar system. Equation $R^3 = \frac{G(M_1 + M_2)}{4\pi^2}T^2$ determines the sum of the masses; and, if R_1 and R_2 are the distances of the individual stars from the centre of mass, the ratio of the distances must balance the inverse ratio of the masses, and the sum of the distances is the total distance R. In symbols:

$$\frac{R_1}{R_2} = \frac{M_2}{M_1};\ R_1 + R_2 = R$$

Those relations are sufficient to determine the individual masses. Observations of the orbital motions of double stars, of the dynamic motions of stars collectively moving within their galaxies, and of the motions of the galaxies themselves verify that Newton's law of gravity is valid to a high degree of accuracy throughout the visible universe.

Ocean tides, phenomena that mystified thinkers for centuries, were also shown by Newton to be a consequence of the universal law of gravitation, although the details of the complicated phenomena were not understood until comparatively recently. They are caused specifically by the gravitational pull of the Moon and, to a lesser extent, of the Sun.

Newton showed that the equatorial bulge of Earth was a consequence of the balance between the centrifugal forces of the rotation of Earth and the attractions of each particle of Earth on all others. The value of gravity at the surface of Earth increases in a corresponding way from the equator to the poles. Among the data that Newton used to estimate the size of the equatorial bulge were the adjustments to his pendulum clock that the English astronomer Edmond Halley had to make in the course of his astronomical observations on the southern island of Saint Helena. Jupiter, which rotates faster than Earth, has a proportionally larger equatorial bulge, the difference between its polar and equatorial radii being about 10 percent. Another success of Newton's theory was his demonstration that comets move in parabolic orbits under the gravitational attraction of the Sun. In a thorough analysis in the Principia, he showed that the great comet of 1680–81 did indeed follow a parabolic path.

It was already known in Newton's day that the Moon does not move in a simple Keplerian orbit. Later, more-accurate observations of the planets also showed discrepancies from Kepler's laws. The motion of the Moon is particularly complex; however, apart from a long-term acceleration due to tides on Earth, the complexities can be accounted for by the gravitational attraction of the Sun and the planets. The gravitational attractions of the planets for each other explain almost all the features of their motions. The exceptions are nonetheless important. Uranus, the seventh planet from the Sun, was observed to undergo variations in its motion that could not be explained by perturbations from Saturn, Jupiter, and the other planets. Two 19th-century astronomers, John Couch Adams of Britain and Urbain-Jean-Joseph Le Verrier of France, independently assumed the presence of an unseen eighth planet that could produce the observed discrepancies. They calculated its position within a degree of where the planet Neptune was discovered in 1846. Measurements of the motion of the innermost planet, Mercury, over an extended period led astronomers to conclude that the major axis of this planet's elliptical orbit precesses in space at a rate 43 arc seconds per century faster than could be accounted for from perturbations of the other planets. In this case, however, no other bodies could be found that could produce this discrepancy, and very slight modification of Newton's law of gravitation seemed to be needed. Einstein's theory of relativity precisely predicts this observed behaviour of Mercury's orbit.

Potential Theory

For irregular, nonspherical mass distributions in three dimensions, Newton's original vector equation $\mathbf{F} = \frac{GM\sum_{1} M_1 \mathbf{r}_1}{r_1^{\,3}}$ is inefficient, though theoretically it could be used for finding the resulting gravitational field. The main progress in classical gravitational theory after Newton was the development of potential theory, which provides the mathematical representation of gravitational fields. It allows practical as well as theoretical investigation of the gravitational variations in space and of the anomalies due to the irregularities and shape deformations of Earth.

Potential theory led to the following elegant formulation: the gravitational acceleration g is a function of position R, g(R), which at any point in space is given from a function Φ called the gravitational potential, by means of a generalization of the operation of differentiation:

$$\mathbf{g}(\mathbf{R}) = \frac{\partial \Phi}{\partial x}\mathbf{i} + \frac{\partial \Phi}{\partial y}\mathbf{j} + \frac{\partial \Phi}{\partial z}\mathbf{k},$$

in which i, j, and k stand for unit basis vectors in a three-dimensional Cartesian coordinate system. The potential and therefore g are determined by an equation discovered by the French mathematician Siméon-Denis Poisson:

$$\left(\frac{\partial^2}{\partial x^2} + \frac{\partial^2}{\partial y^2} + \frac{\partial^2}{\partial z^2}\right)\Phi(\mathbf{R}) = -4\pi G\rho(\mathbf{R}).$$

where ρ(R) is the density at the vector position R.

The significance of this approach is that Poisson's equation can be solved under rather general conditions, which is not the case with Newton's equation. When the mass density ρ is nonzero, the solution is expressed as the definite integral:

$$\Phi(\mathbf{R}) = G\int \frac{\rho(\mathbf{R}')d\mathbf{R}'}{|\mathbf{R} - \mathbf{R}'|}.$$

where the integral is a three-dimensional integral over the volume of all space. When $\rho = 0$ (in particular, outside Earth), Poisson's equation reduces to the simpler equation of Laplace.

The appropriate coordinates for the region outside the nearly spherical Earth are spherical polar coordinates: R, the distance from the centre of Earth; θ, the colatitude measured from the North Pole; and the longitude measured from Greenwich. The solutions are series of powers of R multiplied by trigonometric functions of colatitude and longitude, known as spherical harmonics; the first terms are:

$$\Phi(\mathbf{R}) = \frac{GM_E}{R}\left[1 - J_2\left(\frac{R_E}{R}\right)^2 \frac{3\cos^2\theta - 1}{2} - J_3\left(\frac{R_E}{R}\right)^3 \frac{5\cos^3 - \theta - 3\cos\theta}{2} + \ldots\right]$$

The constants J_2, J_3, and so forth are determined by the detailed mass distribution of Earth; and, since Newton showed that for a spherical body all the J_n are zero, they must measure the deformation of Earth from a spherical shape. J_2 measures the magnitude of Earth's rotational equatorial bulge, J_3 measures a slight pear-shaped deformation of

Earth, and so on. The orbits of spacecraft around Earth, other planets, and the Moon deviate from simple Keplerian ellipses in consequence of the various spherical harmonic terms in the potential. Observations of such deviations were made for the very first artificial spacecraft. The parameters J_2 and J_3 for Earth have been found to be 1,082.7 $\times 10^{-6}$ and -2.4×10^{-6}, respectively. Very many other harmonic terms have been found in that way for Earth and also for the Moon and for other planets. Halley had already pointed out in the 18th century that the motions of the moons of Jupiter are perturbed from simple ellipses by the variation of gravity around Jupiter.

The surface of the oceans, if tides and waves are ignored, is a surface of constant potential of gravity and rotation. If the only spherical harmonic term in gravity were that corresponding to the equatorial bulge, the sea surface would be just a spheroid of revolution (a surface formed by rotating a two-dimensional curve about some axis; for example, rotating an ellipse about its major axis produces an ellipsoid). Additional terms in the potential give rise to departures of the sea surface from that simple form. The actual form may be calculated from the sum of the known harmonic terms, but it is now possible to measure the form of the sea surface itself directly by laser ranging from spacecraft. Whether found indirectly by calculation or directly by measurement, the form of the sea surface may be shown as contours of its deviation from the simple spheroid of revolution.

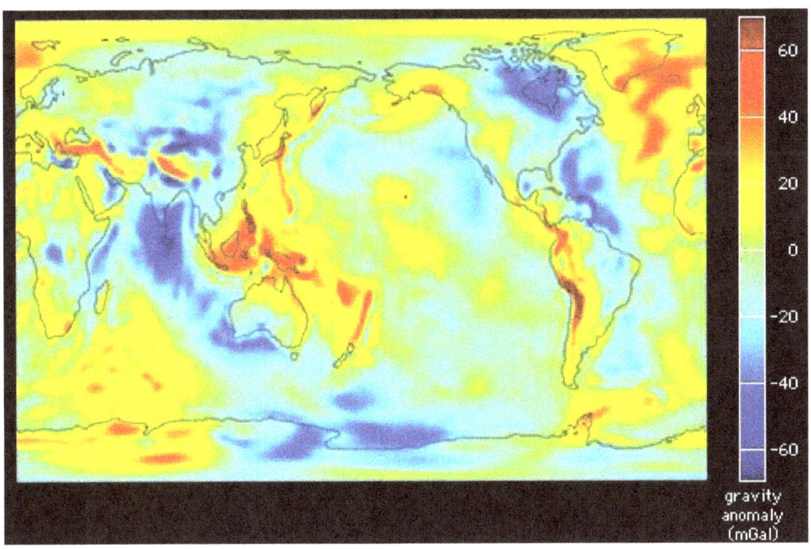

The variation in the gravitational field, given in milliGals (mGal), over the Earth's surface gives rise to an imaginary surface known as the geoid. The geoid expresses the height of an imaginary global ocean not subject to tides, currents, or winds. Such an ocean would vary by up to 200 metres (650 feet) in height because of regional variations in gravitation.

Effects of Local Mass Differences

Spherical harmonics are the natural way of expressing the large-scale variations of potential that arise from the deep structure of Earth. However, spherical harmonics are not suitable for local variations due to more-superficial structures. Not long after Newton's time, it was found that the gravity on top of large mountains is less than expected on the basis of their visible mass. The idea of isostasy was developed, according to which the unexpectedly low acceleration of gravity on a mountain is caused by low-density rock 30 to 100 km underground, which buoys up the mountain. Correspondingly, the unexpectedly high force of gravity on ocean surfaces is explained by dense rock 10 to 30 km beneath the ocean bottom.

Portable gravimeters, which can detect variations of one part in 10^9 in the gravitational force, are in wide use today for mineral and oil prospecting. Unusual underground deposits reveal their presence by producing local gravitational variations.

Weighing the Earth

The mass of Earth can be calculated from its radius and g if G is known. G was measured by the English physicist-chemist Henry Cavendish and other early experimenters, who spoke of their work as "weighing the Earth." The mass of Earth is about 5.98×10^{24} kg, while the mean densities of Earth, the Sun, and the Moon are, respectively, 5.52, 1.43, and 3.3 times that of water.

Acceleration around Earth, the Moon and other Planets

The value of the attraction of gravity or of the potential is determined by the distribution of matter within Earth or some other celestial body. In turn, the distribution of matter determines the shape of the surface on which the potential is constant. Measurements of gravity and the potential are thus essential both to geodesy, which is the study of the shape of Earth, and to geophysics, the study of its internal structure. For geodesy and global geophysics, it is best to measure the potential from the orbits of artificial satellites. Surface measurements of gravity are best for local geophysics, which deals with the structure of mountains and oceans and the search for minerals.

Variations in g

Changes due to Location

The acceleration g varies by about $^1/_2$ of 1 percent with position on Earth's surface, from about 9.78 metres per second per second at the Equator to approximately 9.83 metres per second per second at the poles. In addition to this broad-scale variation, local variations of a few parts in 10^6 or smaller are caused by variations in the density of Earth's crust as well as height above sea level.

Changes with Time

The gravitational potential at the surface of Earth is due mainly to the mass and rotation of Earth, but there are also small contributions from the distant Sun and Moon. As Earth rotates, those small contributions at any one place vary with time, and so the local value of g varies slightly. Those are the diurnal and semidiurnal tidal variations. For most purposes it is necessary to know only the variation of gravity with time at a fixed place or the changes of gravity from place to place; then the tidal variation can be removed. Accordingly, almost all gravity measurements are relative measurements of the differences from place to place or from time to time.

Measurements of g

Unit of Gravity

Because gravity changes are far less than 1 metre per second per second, it is convenient to have a smaller unit for relative measurements. The gal has been adopted for this purpose; a gal is one-hundredth metre per second per second. The unit most commonly used is the milligal, which equals 10^{-5} metre per second per second—i.e., about one-millionth of the average value of g.

Absolute Measurements

Two basic ways of making absolute measurements of gravity have been devised: timing the free fall of an object and timing the motion under gravity of a body constrained in some way, almost always as a pendulum. In 1817 the English physicist Henry Kater, building on the work of the German astronomer Friedrich Wilhelm Bessel, was the first to use a reversible pendulum to make absolute measurements of g. If the periods of swing of a rigid pendulum about two alternative points of support are the same, then the separation of those two points is equal to the length of the equivalent simple pendulum of the same period. By careful construction, Kater was able to measure the separation very accurately. The so-called reversible pendulum was used for absolute measurements of gravity from Kater's day until the 1950s. Since that time, electronic instruments have enabled investigators to measure with high precision the half-second time of free fall of a body (from rest) through one metre. It is also possible to make extremely accurate measurements of position by using interference of light. Consequently, direct measurements of free fall have replaced the pendulum for absolute measurements of gravity.

Nowadays, lasers are the sources of light for interferometers, while the falling object is a retroreflector that returns a beam of light back upon itself. The falling object can be timed in simple downward motion, or it can be projected upward and timed over the upward and downward path. Transportable versions of such apparatuses have been used in different locations to establish a basis for measuring differences of gravity over the entire Earth. The accuracy attainable is about one part in 10^8.

More recently, interferometers using beams of atoms instead of light have given absolute determinations of gravity. Interference takes place between atoms that have been subject to different gravitational potentials and so have different energies and wavelengths. The results are comparable to those from bodies in free fall.

Relative Measurements

From the time of Newton, measurements of differences of gravity (strictly, the ratios of values of gravity) were made by timing the same pendulum at different places. During the 1930s, however, static gravimeters replaced pendulums for local measurements over small ranges of gravity. Today, free-fall measurements have rendered the pendulum obsolete for all purposes.

Spring gravimeters balance the force of gravity on a mass in the gravity field to be measured against the elastic force of the spring. Either the extension of the spring is measured, or a servo system restores it to a constant amount. High sensitivity is achieved through electronic or mechanical means. If a thin wire is stretched by a mass hung from it, the tension in the wire, and therefore the frequency of transverse oscillations, will vary with the force of gravity upon the mass. Such vibrating string gravimeters were originally developed for use in submarines and were later employed by the Apollo 17 astronauts on the Moon to conduct a gravity survey of their landing site. Another relatively recent development is the superconducting gravimeter, an instrument in which the position of a magnetically levitated superconducting sphere is sensed to provide a measure of g. Modern gravimeters may have sensitivities better than 0.005 milligal, the standard deviation of observations in exploration surveys being of the order of 0.01–0.02 milligal.

Differences in gravity measured with gravimeters are obtained in quite arbitrary units—divisions on a graduated dial, for example. The relation between these units and milligals can be determined only by reading the instrument at a number of points where g is known as a result of absolute or relative pendulum measurements. Further, because an instrument will not have a completely linear response, known points must cover the entire range of gravity over which the gravimeter is to be used.

Since g is an acceleration, the problem of its measurement from a vehicle that is moving, and therefore accelerating relative to Earth, raises a number of fundamental problems. Pendulum, vibrating-string, and spring-gravimeter observations have been made from submarines; using gyrostabilized platforms, relative gravity measurements with accuracies approaching a few milligals have been and are being made from surface ships. Experimental measurements with various gravity sensors on fixed-wing aircraft as well as on helicopters have been carried out.

Gravimetric Surveys and Geophysics

As a result of combining all available absolute and relative measurements, it is now possible to obtain the most probable gravity values at a large number of sites to high accuracy. The culmination of gravimetric work begun in the 1960s has been a world-wide gravity reference system having an accuracy of at least one part in 10^7 (0.1 milligal or better).

The value of gravity measured at the terrestrial surface is the result of a combination of factors:

- The gravitational attraction of Earth as a whole.
- Centrifugal force caused by Earth's rotation.
- Elevation.
- Unbalanced attractions caused by surface topography.
- Tidal variations.
- Unbalanced attractions caused by irregularities in underground density distributions.

Most geophysical surveys are aimed at separating out the last of these in order to interpret the geologic structure. It is therefore necessary to make proper allowance for the other factors. The first two factors imply a variation of gravity with latitude that can be calculated for an assumed shape for Earth. The third factor, which is the decrease in gravity with elevation, due to increased distance from the centre of Earth, amounts to −0.3086 milligal per metre. This value, however, assumes that material of zero density occupies the whole space between the point of observation and sea level, and it is therefore termed the free-air correction factor. In practice the mass of rock material that occupies part or all of this space must be considered. In an area where the topography is reasonably flat, this is usually calculated by assuming the presence of an infinite slab of thickness equal to the height of the station h and having an appropriate density σ; its value is +0.04185 σh milligal per metre. This is commonly called the Bouguer correction factor.

Terrain or topographical corrections also can be applied to allow for the attractions due to surface relief if the densities of surface rocks are known. Tidal effects (the amplitudes are less than 0.3 milligal) can be calculated and allowed for.

The Moon and the Planets

Although the Apollo astronauts used a gravimeter at their lunar landing site, most scientific knowledge about the gravitational attractions of the Moon and the planets has been derived from observations of their effects upon the accelerations of spacecraft in orbit around or passing close to them. Radio tracking makes it possible to determine the accelerations of spacecraft very accurately, and the results can be expressed either as terms in a series of spherical harmonics or as the variation of gravity over the surface. As in the case of Earth, spherical harmonics are more effective for studying gross structure, while the variation of gravity is more useful for local features. Spacecraft must descend close to the surface or remain in orbit for extended periods in order to detect local gravity variations; such data had been obtained for the Moon, Venus, Mars, and Jupiter by the end of the 20th century.

The Moon's polar flattening is much less than that of Earth, while its equator is far more elliptical. There are also large, more-local irregularities from visible and concealed structures. Mars also exhibits some large local variations, while the equatorial bulges of Mercury and Venus are very slight.

By contrast, the major planets, all of which rotate quite fast, have large equatorial bulges, and their gravity is dominated by a large increase from equator to pole. The polar flattening of Jupiter is about 10 percent and was first estimated from telescopic observation by Gian Domenico Cassini about 1664. Edmond Halley subsequently realized that the corresponding effect on gravity would perturb the orbits of the satellites of Jupiter. The results of gravity measurements are crucial to understanding the internal properties of the planets.

Gravitational Theory and other aspects of Physical Theory

The Newtonian theory of gravity is based on an assumed force acting between all pairs of bodies—i.e., an action at a distance. When a mass moves, the force acting on other masses had been considered to adjust instantaneously to the new location of the displaced mass. That, however, is inconsistent with special relativity, which is based on the axiom that all knowledge of distant events comes from electromagnetic signals. Physical quantities have to be defined in such a way that certain combinations of them—in particular, distance, time, mass, and momentum—are independent of choice of space-time coordinates. This theory, with the field theory of electrical and magnetic phenomena, has met such empirical success that most modern gravitational theories are constructed as field theories consistent with the principles of special relativity. In a field theory the gravitational force between bodies is formed by a two-step process: (1) One body produces a gravitational field that permeates all surrounding space but has weaker strength farther from its source. A second body in that space is then acted upon by this field and experiences a force. (2) The Newtonian force of reaction is then viewed as the response of the first body to the gravitational field produced by the second body, there being at all points in space a superposition of gravitational fields due to all the bodies in it.

Field Theories of Gravitation

In the 1970s the physicists Abdus Salam of Pakistan and Steven Weinberg and Sheldon L. Glashow of the United States were able to show that the electromagnetic forces and the weak force responsible for beta decay were different manifestations of the same basic interaction. That was the first successful unified field theory. Physicists are actively seeking other possible unified combinations. The possibility that gravitation might be linked with the other forces of nature in a unified theory of forces greatly increased interest in gravitational field theories during the 1970s and '80s. Because the gravitational force is exceedingly weak compared with all others and because it seems to be independent of all physical properties except mass, the unification of gravitation with the other forces remains the most difficult to achieve. That challenge has provided a tremendous impetus to experimental investigations to determine whether there may be some failure of the apparent independence.

The prime example of a field theory is Einstein's general relativity, according to which the acceleration due to gravity is a purely geometric consequence of the properties of space-time in the neighbourhood of attracting masses. In a whole class of more-general theories, these and other effects not predicted by simple Newtonian theory are characterized by free parameters; such formulations are called parameterized post-Newtonian (PPN) theories. There is now considerable experimental and observational evidence for limits to the parameters. So far, no deviation from general relativity has been demonstrated convincingly.

Field theories of gravity predict specific corrections to the Newtonian force law, the corrections taking two basic forms: (1) When matter is in motion, additional gravitational fields (analogous to the magnetic fields produced by moving electric charges) are produced; also, moving bodies interact with gravitational fields in a motion-dependent way. (2) Unlike electromagnetic field theory, in which two or more electric or magnetic fields superimpose by simple addition to give the total fields, in gravitational field theory nonlinear fields proportional to the second and higher powers of the source masses are generated, and gravitational fields proportional to the products of different masses are created. Gravitational fields themselves become sources for additional gravitational fields. Examples of some of these effects are shown below. The acceleration A of a moving particle of negligible mass that interacts with a mass M, which is at rest, is given in the following formula, derived from Einstein's gravitational theory.

The expression for A now has, as well as the Newtonian expression from equation $A = \frac{4\pi^2 R}{T^2}$, further terms in higher powers of GM/R^2—that is, in G^2M^2/R^4. As elsewhere, V is the particle's velocity vector, A is its acceleration vector, R is the vector from the mass M, and c is the speed of light. When written out, the sum is:

$$A = -\frac{GM\mathbf{R}}{R^3} + 2\frac{G^2M^2\mathbf{R}}{c^2R^2} - \frac{3GM\mathbf{R}}{2R^3}\left(\frac{\mathbf{V}^2}{c^2}\right) - \frac{\mathbf{V}\cdot\mathbf{AV}}{c^2} - \frac{1}{2}\frac{\mathbf{V}^2}{c^2}\mathbf{A} + \ldots$$

This expression gives only the first post-Newtonian corrections; terms of higher power in $1/c$ are neglected. For planetary motion in the solar system, the $1/c^2$ terms are smaller than Newton's acceleration term by at least a factor of 10^{-8}, but some of the consequences of these correction terms are measurable and important tests of Einstein's theory. It should be pointed out that prediction of new observable gravitational effects requires particular care; Einstein's pioneer work in gravity has shown that gravitational fields affect the basic measuring instruments of experimental physics—clocks, rulers, light rays—with which any experimental result in physics is established. Some of these effects are listed below:

- The rate at which clocks run is reduced by proximity of massive bodies; i.e., clocks near the Sun will run slowly compared with identical clocks farther away from it.
- In the presence of gravitational fields, the spatial structure of physical objects is no longer describable precisely by Euclidean geometry; for example, in the arrangement of three rigid rulers to form a triangle, the sum of the subtended angles will not equal 180°. A more-general type of geometry, Riemannian geometry, seems required to describe the spatial structure of matter in the presence of gravitational fields.
- Light rays do not travel in straight lines, the rays being deflected by gravitational fields. To distant observers the light-propagation speed is observed to be reduced near massive bodies.

Gravitational Fields and the Theory of General Relativity

In Einstein's theory of general relativity, the physical consequences of gravitational fields are stated in the following way. Space-time is a four-dimensional non-Euclidean continuum, and the curvature of the Riemannian geometry of space-time is produced by or related to the distribution of matter in the world. Particles and light rays travel along the geodesics (shortest paths) of this four-dimensional geometric world.

There are two principal consequences of the geometric view of gravitation: (1) the accelerations of bodies depend only on their masses and not on their chemical or nuclear constitution, and (2) the path of a body or of light in the neighbourhood of a massive body (the Sun, for example) is slightly different from that predicted by Newton's theory. The first is the weak principle of equivalence. Newton himself performed experiments with pendulums that demonstrated the principle to better than one part in 1,000 for a variety of materials, and, at the beginning of the 20th century, the Hungarian physicist Roland, Baron von Eötvös, showed that different materials accelerate in Earth's field at the same rate to within one part in 10^9. More-recent experiments have shown the equality of accelerations in the field of the Sun to within one part in 10^{11}. Newtonian theory is in accord with these results because of the postulate that gravitational force is proportional to a body's mass.

Inertial mass is a mass parameter giving the inertial resistance to acceleration of the body when responding to all types of force. Gravitational mass is determined by the strength of the gravitational force experienced by the body when in the gravitational field g. The Eötvös experiments therefore show that the ratio of gravitational and inertial mass is the same for different substances.

In Einstein's theory of special relativity, inertial mass is a manifestation of all the forms of energy in a body, according to his fundamental relationship $E = mc^2$, E being the total energy content of a body, m the inertial mass of the body, and c the speed of light. Dealing with gravitation, then, as a field phenomenon, the weak principle of equivalence indicates that all forms of nongravitational energy must identically couple to or interact with the gravitational field, because the various materials in nature possess different fractional amounts of nuclear, electrical, magnetic, and kinetic energies, yet they accelerate at identical rates.

In the theory of general relativity, the gravitational field also interacts with gravitational energy in the same manner as with other forms of energy, an example of that theory's universality not possessed by most other theories of gravitation.

The Sun has an appreciable fraction of internal gravitational energy, and the repetitions of the Eötvös experiments during the 1970s, with the Sun instead of Earth as the attracting mass, revealed that bodies accelerate at identical rates in the Sun's field as well as in that of Earth. Extremely accurate laser measurements of the distance of the Moon from Earth have made possible a further test of the weak principle of equivalence. The chemical constitutions of Earth and the Moon are not the same, and so, if the principle did not hold, they might accelerate at different rates under the Sun's attraction. No such effect has been detected.

Newton's third law of dynamics states that every force implies an equal and opposite reaction force. Modern field theories of force contain this principle by requiring every entity that is acted upon by a field to be also a source of the field. An experiment by the American physicist Lloyd Kreuzer established to within 1 part in 20,000 that different materials produce gravitational fields with a strength the same as that of gravitational fields acting upon them. In this experiment a sphere of solid material was moved through a liquid of identical weight density. The absence of a gravitational effect on a nearby Cavendish balance instrument during the sphere's motion is interpreted as showing that the two materials had equal potency in producing a local gravitational-field anomaly.

Other experiments have confirmed Einstein's predictions to within a few percent. Using the Mössbauer effect to monitor the nuclear reabsorption of resonant gamma radiation, a shift of wavelength of the radiation that traveled vertically tens of metres in Earth's gravitational field was measured, and the slowing of clocks (in this case the nuclear vibrations are clocks) as predicted by Einstein was confirmed to 1 percent precision. If v and Δv are clock frequency and change of frequency, respectively, h is the height difference between clocks in the gravitational field g. Then:

$$\frac{\Delta v}{v} = -\frac{gh}{c^{2}}.$$

Paths of Particles and Light

The idea that light should be deflected by passing close to a massive body had been suggested by the British astronomer and geologist John Michell in the 18th century. However, Einstein's general relativity theory predicted twice as much deflection as Newtonian physics. Quick confirmation of Einstein's result came from measuring the direction of a star close to the Sun's direction during an expedition led by the British astronomer Sir Arthur Stanley Eddington to observe the solar eclipse of 1919. Optical determinations of the change of direction of a star are subject to many systematic errors, and far better results have been obtained of the directions of spacecraft with radio interferometers of very long baselines. The effect comes from the decrease in the speed of light near a massive object (the Sun). That decrease has also been found directly from the round-trip travel times for radar pulses between Earth and other inner planets or artificial satellites passing behind the Sun and has confirmed to about 4 percent the prediction of an additional time delay Δt given by the following formula, in which M_S is the Sun's mass, R_1 and R_2 are the distances from the Sun to Earth and to the other reflecting body, and D is the distance of closest approach of the radar pulses to the Sun:

$$\Delta t = \frac{4GM_S}{c^3} \ln \frac{4R_1R_2}{D^2}$$

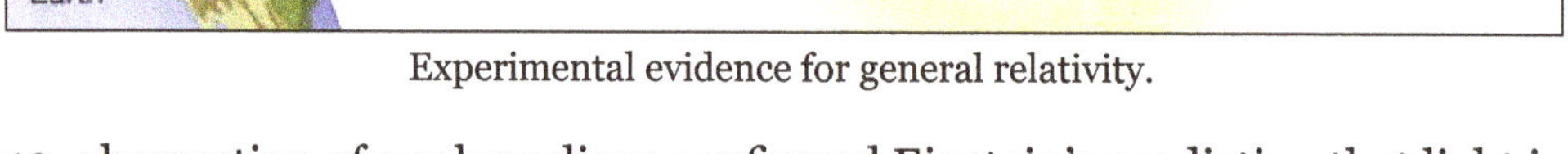

Experimental evidence for general relativity.

In 1919, observation of a solar eclipse confirmed Einstein's prediction that light is bent in the presence of mass. This experimental support for his general theory of relativity garnered him instant worldwide acclaim.

The additional precession of the orbit of Mercury of 43 arc seconds per century was known before the development of the theory of general relativity. With radar measurements of the distances to the planets, similar anomalous precessions have been estimated for Venus and Earth and have been found to agree with general relativity.

Gravitational Radiation

According to general relativity, the curvature of space-time is determined by the distribution of masses, while the motion of masses is determined by the curvature. In consequence, variations of the gravitational field should be transmitted from place to place as waves, just as variations of an electromagnetic field travel as waves. If the masses that are the source of a field change with time, they should radiate energy as waves of curvature of the field. There are strong grounds for believing that such radiation exists. One particular double-star system has a pulsar as one of its components, and, from measurements of the shift of the pulsar frequency due to the Doppler effect, precise estimates of the period of the orbit show that the period is changing, corresponding to a decrease in the energy of the orbital motion. Gravitational radiation is the only known means by which that could happen.

Double stars in their regular motions (such as that for which a change in period has been detected) and massive stars collapsing as supernovas have been suggested as sources of gravitational radiation, and considerable theoretical effort has gone into calculating the signals to be expected from those and other sources.

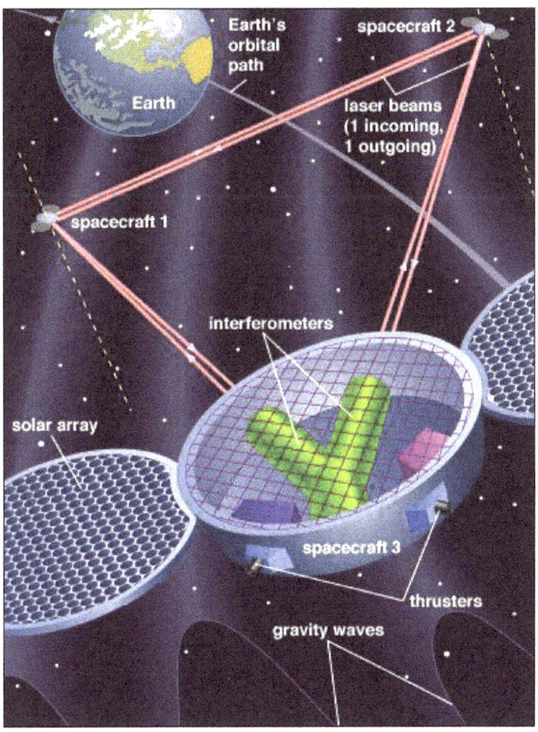

Laser Interferometer Space Antenna (LISA).

LISA, a Beyond Einstein Great Observatory, is scheduled for launch in 2015. Jointly funded by the National Aeronautics and Space Administration (NASA) and the European Space Agency (ESA), LISA will consist of three identical spacecraft that will trail the Earth in its orbit by about 50 million km (30 million miles). The spacecraft will contain thrusters for maneuvering them into an equilateral triangle, with sides of approximately 5 million km (3 million miles), such that the triangle's centre will be located along the Earth's orbit. By measuring the transmission of laser signals between the spacecraft (essentially a giant Michelson interferometer in space), scientists hope to detect and accurately measure gravity waves.

Gravitational radiation is very weak. The changes of curvature would correspond to a dilation in one direction and a contraction at right angles to that direction. One scheme, first tried out about 1960, employed a massive cylinder that might be set in mechanical oscillation by a gravitational signal. The authors of this apparatus argued that signals had been detected, but their claim was not substantiated. In a more fruitful scheme an optical interferometer is set up with freely suspended reflectors at the ends of long paths that are at right angles to each other. Shifts of interference fringes corresponding to an increase in length of one arm and a decrease in the other would indicate the passage of gravitational waves. Such an interferometer, the Laser Interferometer Gravitational-Wave Observatory (LIGO), made the first detection of gravitational radiation in 2015. Two black holes about 1.3 billion light-years away spiralled into each other. The black holes were 36 and 29 times the mass of the Sun and formed a new black hole 62 times the mass of the Sun. In the merger, three solar masses were converted to energy in gravitational waves; the amount of power radiated was 50 times more than that of all the stars shining in the universe in that moment. As of 2019, LIGO has made 10 detections of gravitational radiation. Nine were from mergers of a black hole binary, and one was the merger of a neutron star binary.

Some Astronomical aspects of Gravitation

Studies of gravity allow the masses and densities of celestial bodies to be estimated and thereby make it possible to investigate the physical constitutions of stars and planets. Because gravitation is a very weak force, however, its distinctive effects appear only when masses are extremely large. The idea that light might be attracted gravitationally had been suggested by Michell and examined by the French mathematician and astronomer Pierre-Simon Laplace. Predictions by classical physics and general relativity that light passing close to the Sun might be deflected are described above. There are two further consequences for astronomy. Light from a distant object may pass close to objects other than the Sun and be deflected by them. In particular, they may be deflected by a massive galaxy. If some object is behind a massive galaxy, as seen from Earth, deflected light may reach Earth by more than one path. Operating like a lens that focuses light along different paths, the gravity of the galaxy may make the object appear multiple; examples of such apparently double objects have been found.

Both Michell and Laplace pointed out that the attraction of a very dense object upon light might be so great that the light could never escape from the object, rendering it invisible. Such a phenomenon is a black hole. The relativistic theory of black holes has been thoroughly developed in recent years, and astronomers have conducted extensive observations of them. One possible class of black holes comprises very large stars that have used up all of their nuclear energy so that they are no longer held up by radiation pressure and have collapsed into black holes (less-massive stars may collapse into neutron stars). Supermassive black holes with masses millions to billions times that of the Sun are thought to exist at the centres of most galaxies.

Black holes, from which no radiation is able to escape, cannot be seen by their own light, but there are observable secondary effects. If a black hole were one component of a double star, the orbital motion of the pair and the mass of the invisible member might be derived from the oscillatory motion of a visible companion. Because black holes attract matter, any gas in the vicinity of an object of this kind would fall into it and acquire, before vanishing into the hole, a high velocity and consequently a high temperature. The gas may become hot enough to produce X-rays and gamma rays from around the hole. Such a mechanism is the origin of at least some powerful X-ray and radio astronomical sources, including those at the centres of galaxies and quasars. In the case of the massive galaxy M87, the supermassive black hole at its centre, which has a mass 6.5 billion times that of the Sun has been directly observed.

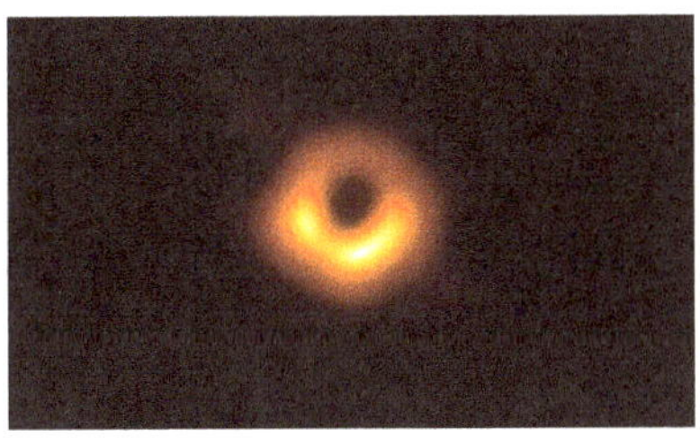

Black hole in M87.

Black hole at the centre of the massive galaxy M87, about 55 million light-years from Earth, as imaged by the Event Horizon Telescope (EHT). The black hole is 6.5 billion times more massive than the Sun. This image was the first direct visual evidence of a supermassive black hole and its shadow. The ring is brighter on one side because the black hole is rotating, and thus material on the side of the black hole turning toward Earth has its emission boosted by the Doppler effect. The shadow of the black hole is about five and a half times larger than the event horizon, the boundary marking the black hole's limits, where the escape velocity is equal to the speed of light.

Gravimetry

Gravimetry deals with the gravity of the earth. It belongs to geodesy as well as to geophysics. While geodesy deals with questions of the earth's figure, especially the determination of the earth's outer gravity field and the setting-up of a global network of measured points as a reference system, geophysics takes more interest in the density distribution in the earth's interior. A measured deviation from the 'normal' distribution of gravity allows to draw conclusions on density inhomogeneities in the underground.

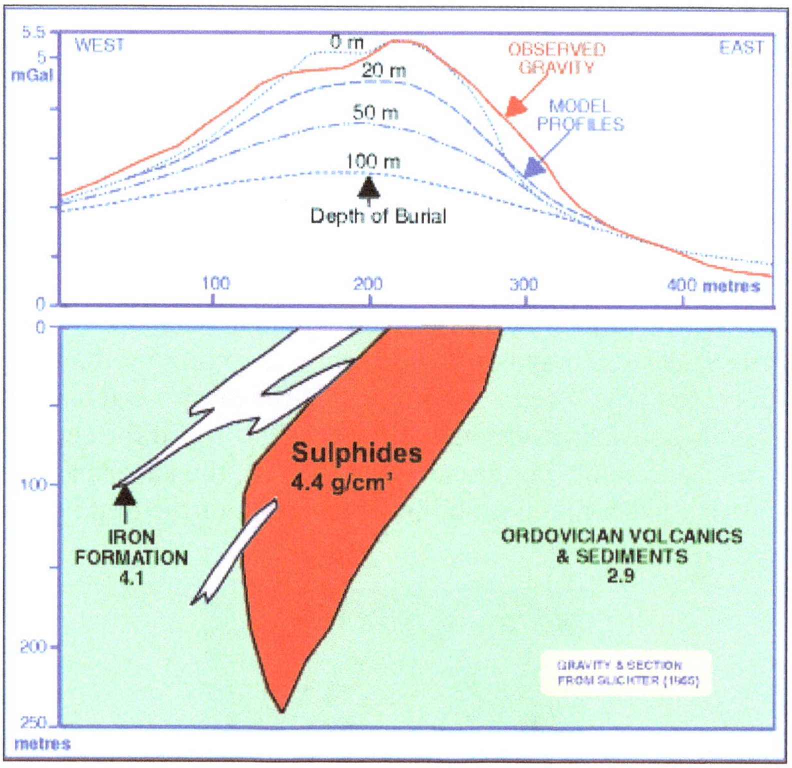

Adequate mathematical modelling methods allow quantitative statements on geometry and density contrasts of the disturbing bodies. Those can be, for example, magmatic intrusive bodies, lithologic boundaries, sedimentary basins, changes of crust-thickness or inhomogeneities or other causes within the earth's crust or the upper mantle. In applied geophysics, gravimetry is applied for the exploration of near-surface layers, for example in hydrocarbon exploration, in the exploration of deposits of metallic (example of a sulphide ore body in the figure) and non-metallic raw materials, in hydrogeology, in foundation soil and environmental surveys.

Gravity is measured with gravimeters. It can be determined absolutely – for example with the help of pendulums in fall experiments or today with so-called superconducting gravimeters.

Such measurements, however, are always very time-consuming in order to achieve adequate accuracy. In practice, measurements are normally carried out relatively, i. e. gravity changes between two measuring points are determined. Such gravimeters consist, in principle, of a small sample mass suspended on an extremely sensitive spring system. The measuring accuracy of modern gravimeters is about 10^{-8} to 10^{-9} of the overall strength of gravity at the earth's surface. Gravimeters are as sensitive as the mechanical possibilities allow.

Marine Gravimetry

Marine gravimetry came to existence as a standard gravimetric technique some 50 years ago, even though the first attempts were made even earlier, in the 1920s and 1930s, on board submarines. It was clear from the beginning that marine measurements could not compete with land measurements in terms of precision. At sea, the gravimeter is subjected to the accelerations of the vessel, whose effects on gravimetry must be addressed. Any acceleration a experienced 'by the vessel' at the location of the gravimeter is felt also by gravimeter aboard the vessel. This acceleration a is automatically added to the real gravity vector, g, so that the meter records the sum of the values:

$$\mathbf{g}^* = \mathbf{g} + \mathbf{a}$$

The geometry of this relationship is shown in figure. This must be countered by modeling and removing, as well as possible, the effects of the vessel acceleration. The most important component of a is the vertical acceleration, which has a full impact on the recorded g* and is the main source of error in marine gravity measurement. Marine gravimetry accuracy is at best on the order of 3–500 μGal.

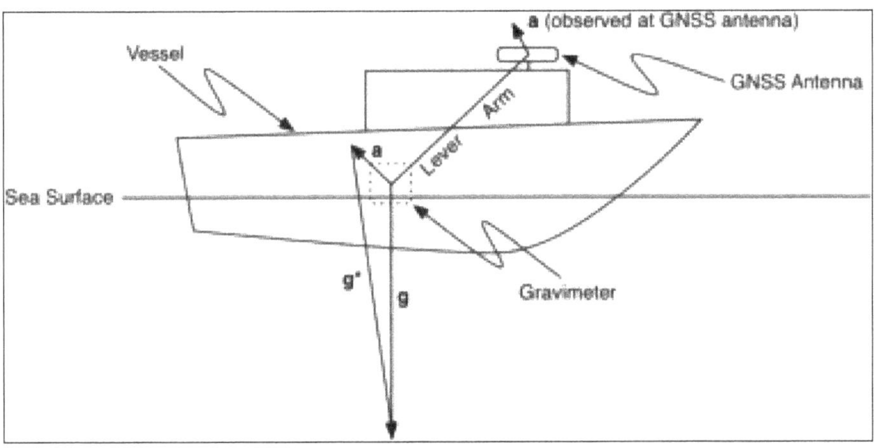

Nowadays, three-dimensional vessel acceleration can be measured quite well using multi-antenna global navigation satellite systems(GNSS), used normally in conjunction with an inertial measurement unit (IMU) comprising an array of accelerometers and gyroscopes. Because the GNSS antennas and the IMU are not located at the same place as the gravimeter, they may be subject to slightly different accelerations, and these differences must be accounted for in what is called a lever arm correction.

Wave-induced vessel motions subject a gravimeter at sea to high-frequency (above 0.1 Hz) accelerations with magnitudes that may exceed 10 Gal. The resulting incessant beam motion precludes measurement by adjusting spring tension, as in land gravimetry. To counteract this problem, the beam motion of spring gravimeters for use at sea is heavily dumped (of near-infinite period). This reduces effects of vessel motion and more importantly allows reading to a reasonable accuracy based on beam velocity. The spring still plays a role in the measurement as it is being adjusted to keep beam oscillations within a reasonable range. Effects of vessel motion that act on the spring despite gravimeter damping are periodic with short periods and in marine gravimetry are effectively removed in postprocessing by filtering over a 3–5 min window using a finite impulse response filter.

It is customary to further minimize the influence of vessel motions by mounting the gravimeter on a stabilized platform. The platform's orientation is continuously adjusted by servomotors, controlled by gyroscopes and accelerometers, in an attempt to preserve the vertical orientation of the gravimeter. A simplified diagram of a stabilized platform is shown in figure.

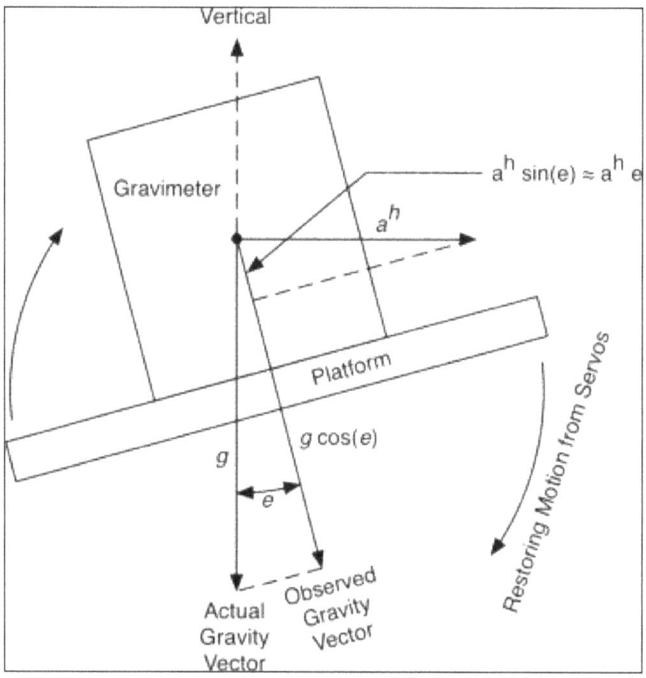

Stabilized gravimeter platform.

The platform stabilization is imperfect, and as a result, two additional corrections must be applied to the observed gravity. One is called the off-level or platform tilt correction and is required because the gravimeter only measures the component of gravity in the direction in which it is oriented. If the tilt angle, e, and the horizontal acceleration, a^h, in the tilt direction are recorded by the platform, the tilt correction is applied:

$$\delta g^t = -a^h - ge^2/2$$

This correction converts the component of gravity observed by the off-level gravimeter to the correct magnitude of the gravity vector.

The second error induced by nonvertical motions is known as cross coupling and results from couples (rotations) in the gravimeter beam that arise due to vessel motions.

The mechanisms by which this happens and the methods of correcting it rely on mathematical modeling of the effects and are too complicated to treat here. The cross coupling and platform tilt errors result from unintentional motion of the vessel.

A more significant effect of horizontal vessel motion that acts on the gravimeter regardless of stabilization is the Coriolis force, an additional inertial force experienced by a physical body that moves with respect to the spinning Earth. The corresponding Coriolis' acceleration, a_C, is given by the formula:

$$a_C = -2\Omega \times v$$

where Ω denotes is the rotation vector, vis the velocity of the moving body with respect to the Earth and $\times$ denotes the vector cross product.

The vertical component of the Coriolis acceleration is sensed by moving gravimeters and must be removed from observed gravity. This effect was first formulated by Hungarian physicist Loránd Eötvös and is therefore called the Eötvös effect. It is routinely removed from observations using the Eötvös correction, which can be calculated with sufficient accuracy for marine gravimetry using the formula:

$$\delta g^E = 2\Omega v_{EW} \cos\varphi + \frac{|v|^2}{R}$$

where v_{EW} is the east–west horizontal component of v, φ is the latitude of the vessel, and R is the mean radius of the Earth. The first term accounts for the difference in centrifugal acceleration due to the vessel's east–west velocity relative to a rotating, Earth-fixed coordinate system. The second term accounts for the centrifugal acceleration resulting from the motion of the vessel along the curved surface of the Earth. For the latitude of 45°, the bulk of the correction is 0.455 mGal times the east–west velocity component in kilometer per hour.

With marine gravimetry, any required ties to a (national) gravimetric network have to be along the shore line and must be made while the vessel is docked. These are made at the start and end of a cruise and are analogous to the ties used in land gravity surveying. As the ship sails further from the shore, nonlinear gravimeter drift and other unmodeled effects will make the observed gravity less precise. An effective countermeasure is the so-called crossover point adjustment, which requires that gravity survey lines cross each other regularly, so that the whole survey is in the form of a two-dimensional mesh. The gravity values at the crossover points along each of the crossing survey lines are interpolated, as illustrated in figure, and the difference between these two values provides an error estimate at each crossover point known as the crossover error. On the basis of these errors, error model (usually only linear) parameters are estimated so as to minimize the crossover errors. The crossover differences are themselves a good measure of the internal consistency of the survey and are a commonly used indicator of the survey precision.

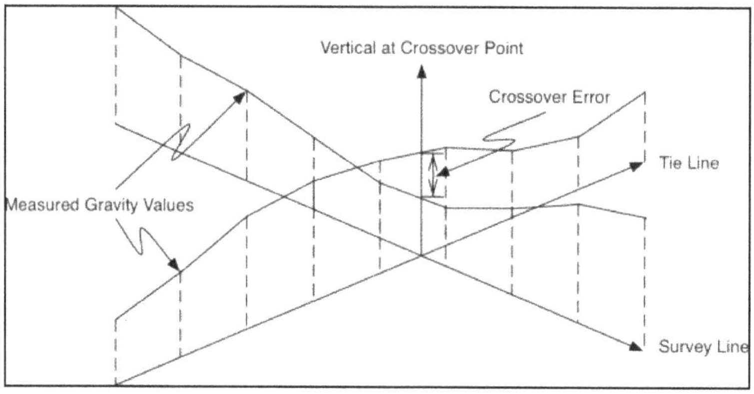

Crossover error.

Turns of the vessel tend to register with the gravimeter despite all the countermeasures, and their influences persist for some time after completing the turn. For this reason, readings taken 6–10 min (called the 'lead-in time') after the turn are ignored.

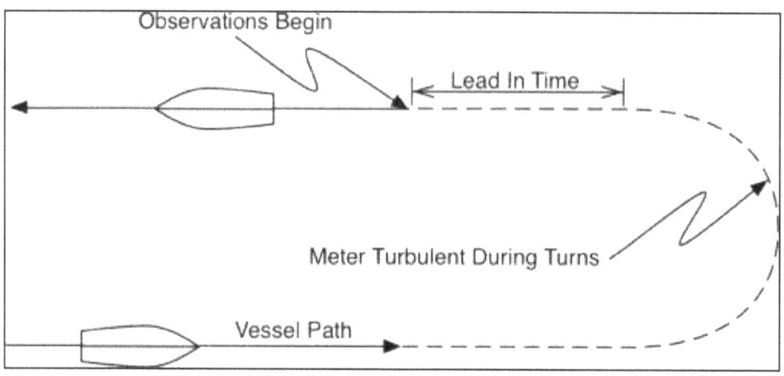

Lead-in time.

A further source of information on marine gravity anomalies rather than on gravity itself, is satellite altimetry. In this technique, orbiting satellites with precisely determined orbits use radar distance measurement to determine the height of the ocean surface, which after filtering out extraneous effects gives approximate geoid heights. The geoid heights can be then converted to deflections of the vertical and to gravity anomalies. The mathematical formulation of the transformation can be found in Sandwell and Smith. Once the gravity anomalies $\Delta g(\varphi, \lambda)$ are obtained for all desired positions (φ, λ), gravity can be derived from them using a very simple formula:

$$g(\phi,\lambda)=\gamma(\phi)+\Delta g(\phi,\lambda)$$

where $\gamma(\phi)$ is the normal gravity, usually the one accepted by the International Association of Geodesy in 1981. The overall procedure amounts to a high-pass filtering of the smooth geoid to calculate the rough gravity anomalies and requires many approximations. As a result, gravity anomalies from satellite altimetry are of worse quality than other data sources, having accuracies on the order of 2–20 mGal, with worse errors when the seafloor is rough or the sea surface is disturbed. Altimetry is also not usable near coasts, because the sea surface is disturbed by a variety of shore-related influences that cannot be removed. Despite these problems, satellite altimetry is a useful data source in many areas where other observations are sparse or nonexistent.

It should be mentioned that gravity is being measured also at the bottom of seas and lakes. The gravimeters used for this purpose have to be specially designed, and the observations corrected for the effect of the varying water column above the instrument. The attraction of the tidal increment is computed from tidal maps.

Gravimeter

Gravimeters are used when prospecting for subterranean deposits of valuable natural resources, including petroleum and minerals, as well as by geodesists to study the shape of the earth and its gravitational field. All modern gravimeters are designed with a highly stable inertial platform to balance the masking effects of motion and vibration.

Types of Gravimeters

Modern land gravimeters are broadly classified into two types – absolute and relative.

Absolute Free-fall Gravimeters

Absolute gravimeters measure the rate at which a mass in free fall in a vacuum accelerates. A retroflector is used with a Michelson interferometer to count and time interference fringes and thus measure the acceleration of the mass.

The types of absolute free-fall gravimeters are:

- Falling corner-cube gravimeter: The motion of a free falling corner-cube retroreflector in vacuum is monitored by a laser interferometer, which detects optical interference to determine the rate of acceleration of gravity. Corner-cube absolute free-fall gravimeters are one of the most accurate types; however, their mechanical structure for repeated free-falling is not not suitable for mobile use and restricts their cycle time.

- Atom interferometer gravimeter: The acceleration of laser-cooled particles due to gravity is determined by calculating the phase difference of a wave packet of particles that is divided through a beamsplitter, with each beam travelling along two different paths, one slightly more elevated than the other and so experiencing slightly different gravity, before being recombined. This technique benefits from high sensitivity and accuracy, and fast cycle times. It is the most suitable gravimeter for mobile use.

Spring-based Relative Gravimeters

Spring-based gravimeters measure only relative gravity by analysing the response of a spring to an attached mass acting in a gravitational field. They are often used in gravity surveys over large areas as they are compact, transportable systems with good sensitivity and drift rate.

The types of spring-based gravimeters are a spring gravimeter has good sensitivity even on a mobile platform, but it needs regular recalibration and has the limitations of a mechanical spring, such as aging. An ideal-spring gravimeter has the best sensitivity relative to the other gravimeters. However, the ideal-spring gravimeter is limited by the environment, and it is difficult to use on a mobile platform.

Spring Gravimeter (LaCoste-Romberg Gravimeter)

Using a zero-length spring, we can implement a LaCoste-Romberg gravimeter. A zero-length spring has no extension corresponding to zero initial force. This spring uses a linearized stress-strain curve around the origin; the length of the spring is directly proportional to the external force. The acceleration of gravity, g is measured at the equilibrium state. The temperature-controlled zero-length spring sustains a mass against gravity, and gravity variations change the spring's length. Important errors arise from the cross coupling between horizontal and vertical accelerations and a gravimeter's imperfections. Most of these errors can be corrected by using a dynamically stabilized platform. The platform provides highly accurate gravity measurements from a moving ship or aircraft. A spring gravimeter is a compact transportable system that has good sensitivity (1×10^{-10} $gHz^{-1/2}$) and drift rate ($\delta_g/g = 3 \times 10^{-8}$ day^{-1}). The non-ideal mechanical spring always needs recalibration, and most drift comes from the aging of the spring, which changes the spring constant.

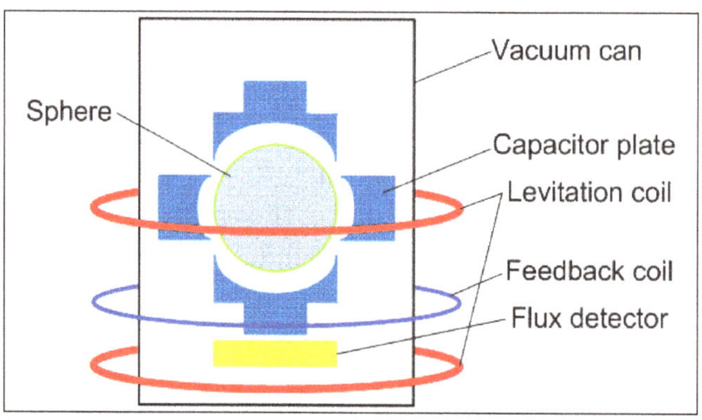

Diagram of the cryogenic portion of the Ideal-spring Gravimeter.

Ideal-spring Gravimeter (Superconducting Gravimeter)

A superconducting sphere is the best candidate for replacing a mechanical spring with an ideal spring. The sensitivity is up to 10^{-12} g $Hz^{-1/2}$, and the drift rate is about $\delta_g/g = 2$

x 10^{-10} day^{-1}. Superconductivity occurs when the superconducting material goes below a sufficiently low temperature, such as a few degrees Kelvin. Below the critical temperature, currents move through the material without dissipation. Depending on the zero resistance property of superconductors, the newly generated magnetic field from a persistent current causes an applied external magnetic field to be excluded from the inside of the sphere. The magnetic levitation force, which can lift the superconducting sphere, comes from the interaction between the inhomogeneous magnetic field of the coils and the persistent currents induced by it in the superconducting sphere. On the vertical axis, there are two Helmholtz-type coils which levitate the superconducting sphere above the plane of the upper coil. The superconducting sphere trapped in the inhomogeneous magnetic field is working as the ideal spring because of repulsive magnetic force. The ideal-spring superconducting gravimeter has perfect stability, which comes from the stability of the persistent currents.

The position of the sphere is measured by a magnetic flux detector and six capacitor plates. A magnetic flux detector at the bottom of the chamber detects variation of the vertical magnetic field. The measured capacitance between the sphere and the capacitor plates provides the precise position of the sphere. Based on the measured position of the sphere, the total levitating force and the force gradient can be adjusted by the magnetic levitation. We can control the position of the sphere and the restoring force on it. Hence, the sphere can be supported and be made responsive to a small change in gravity.

The sensitivity of a gravimeter composed of a superconducting sphere is expected to be much higher than that of a spring gravimeter. The performance of the ideal-spring gravimeter on a moving platform is limited even though the variations of the sphere in gravity are counter-balanced using force feedback. When the instrument is moved, vibration-induced magnetic flux-jumps in the superconductor cause gravity offsets in the gravity reading.

Absolute Free-fall Gravimeters

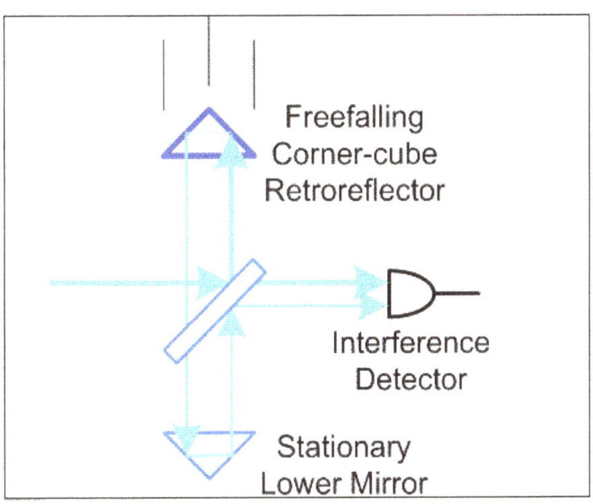

Falling Corner-cube Gravimeter.

Falling corner-cube gravimeters and atom interferometer gravimeters are absolute free-fall gravimeters. These absolute free-fall gravimeters can also be used for measuring the gravitational constant G. The corner-cube absolute free-fall gravimeter has the best accuracy of the gravimeters, but its mechanical structure for repeated free-falling prevents its use on a mobile platform and limits its cycle time. The atom interferometer gravimeter has good accuracy with a fast cycle time and a high mobility relative to other gravimeters.

Falling Corner-cube Gravimeter

A laser interferometer monitors the motion of a freely falling corner-cube retroreflector in a vacuum. The interferometer detects optical fringes which come from optical interference. This signal is used to determine the local acceleration of gravity. This falling corner-cube gravimeter is an absolute free-fall gravimeter. When the drop rate is about 1/15 Hz, the sensitivity is 5×10^{-8} g/$Hz^{-1/2}$ and the accuracy is $\delta g/g = 2 \times 10^{-9}$. As the drop rate becomes higher, vibration-induced accuracy degradation occurs. The sensitivity and accuracy strongly depends on place and drop rate. The sensitivity and accuracy are comparable to those in the atom interferometer gravimeter, but the falling corner-cube gravimeter has some limitations in terms of mobility due to its mechanical dropping platform.

Atom Interferometer Gravimeter

An atom interferometer gravimeter with a sequential light pulse can be implemented by an atomic-fountain interferometer. It has good sensitivity and good accuracy (about 4.2×10^{-9} g/$Hz^{-1/2}$ and $\Delta g\ /g = 3 \times 10^{-9}$. It also has fast cycle time (1.3 sec). Laser-cooled wave packets of cold atoms are sequentially exposed to $\pi/2$-π- π /2 pulse beams. The principle is similar to that of an optical interferometer. For deBroglie matter waves, i.e., cold atoms, a $\pi/2$ pulse works as a beam splitter, and a π pulse works as a reflector. Cold atoms with an initial state are prepared at the bottom of the chamber, and the atoms are launched upward. The first $\pi/2$ pulse at t_0 creates a superposition of states |1> and |2>, and the atomic beams are split as shown in position 1. The atoms are almost stopped at the top of the chamber because of gravity. The second π pulse beams at $t_0 + T$ deflect each partial atomic beam in opposite directions at positions 2 and 3. The atoms fall towards the starting position. The third $\pi/2$ pulse beams at $t_0 + 2T$ recombine the two partial atomic beams, which causes the atomic wave packet to interfere with the other states' atoms as shown in position 4. The atomic phase difference along the different paths allows the determination of g, the acceleration due to gravity. In addition, the limited stabilization of laser frequency and phase limit the precision of atom interferometer gravimeter.

Performance Comparison

A spring gravimeter has good sensitivity and good transportability, but it has the limits of a mechanical spring and low cycle time. An ideal-spring gravimeter overcomes the problem of a spring gravimeter allowing greater sensitivity, but it is difficult to use the gravimeter on a moving platform. In addition, both gravimeters are spring-based relative gravimeters which cannot measure the absolute value of gravity; the performance factor, accuracy, is applicable only to absolute free-fall gravimeters.

A falling corner-cube gravimeter and an atom interferometer gravimeter have good sensitivity and accuracy, and both are absolute free-fall gravimeters. A falling corner-cube gravimeter has limited mobility and low cycle time due to its repeatedly falling retroreflector. The performance of an atom interferometer gravimeter is limited by the limited stabilization of laser frequency and phase, but it has good.

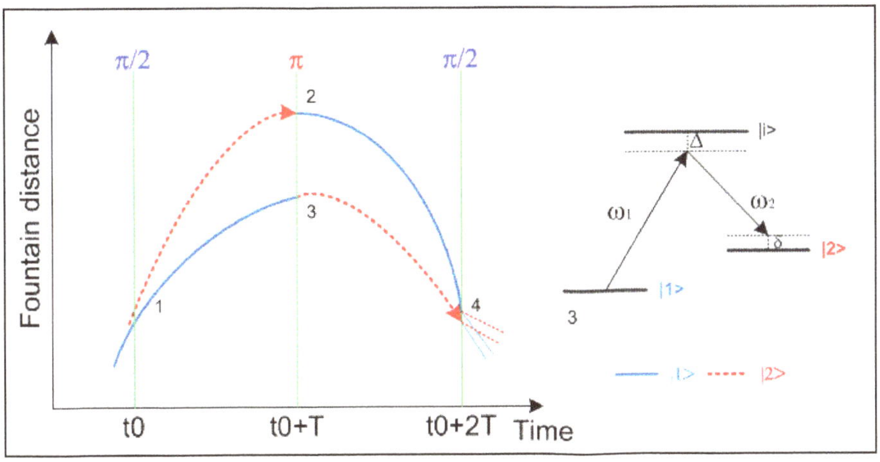

Light-pulse atomic interferometer.

Earth's Radioactivity

Our planet is warm. The pleasant conditions we enjoy on its surface are principally due to the radioactive processes taking place at its centre. The Earth radioactivity causes our planet to behave like an immense hot-water bottle: slowing down the cooling rate and consequently making it habitable. A small half of the heat necessary for our survival is released by the radioactive disintegrations which take place in the rocks that form our Earth crust. The other half of the released power comes mainly from the very slow cooling of the Earth since its formation, which is called the secular cooling.

The blue planet.

Radioactivity has played a major role in giving the Earth its blue and friendly appearance, by preventing it from cooling down. Together with the heat produced by the Sun, it has also guaranteed the pleasant temperatures necessary for life as we know it.

The thermal energy emerging from the ground has been estimated to be around 46 trillion watts (46 TW or terawatts), including 2 TW from volcanic eruptions and earthquakes. Apart from earthquakes and volcanic eruptions, most of this geotherma flux is released in a continuous way. This is called the geothermal flow.

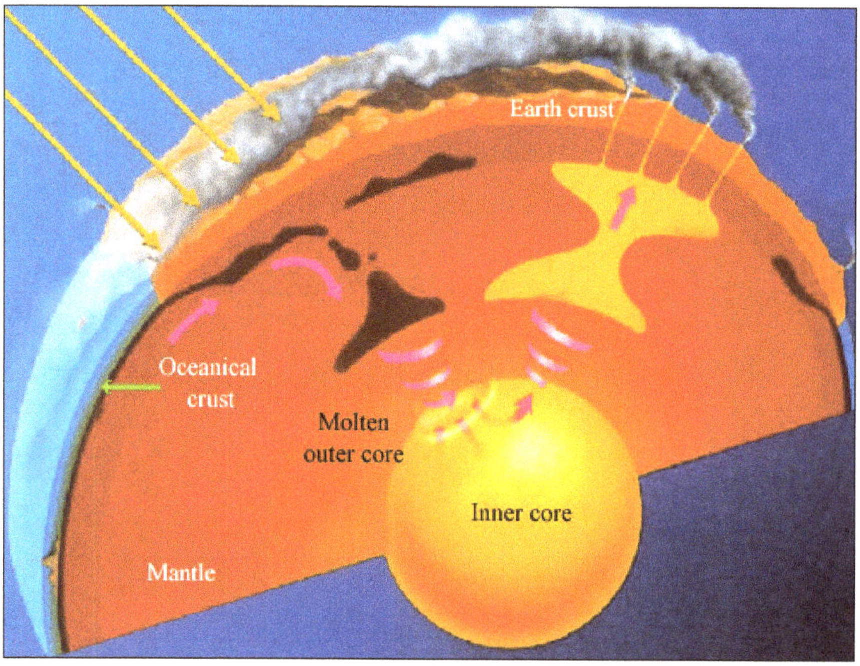

A permanent source of heat.

The bowels of the Earth closely resemble a giant thermal power station, powered by convection currents that heat the surface by taking energy from the rocks buried deeper underground. At the very centre of this giant factory we have the inner core and the molten outer core. Moving outwards, we have the terrestrial mantle, followed by the Earth's crust. The outer layer of the mantle and the whole of the crust are sometimes collectively referred to as the lithosphere: a puzzle comprised of twelve enormous pieces that move against each other to the rhythm of powerful subterranean currents. These pieces are known as tectonic plates, and the boundaries between them are where the Earth's inner workings can most clearly be seen; as these are places where one frequently finds earthquakes and volcanoes.

A part of this geothermal flow - between 15 and 25 TW - is due to a small but constant heat release resulting from the disintegration of long lived radioactive atoms present in the Earth. Owing to the extremely long lifetimes (several billions of years) of surviving radioactive atoms, the heat releases are very small : very small, since the heat flux amounts around 0,1 watt/m^2. it would take many tons of earth merely to power a light bulb. Quasi constant, since this release has only halved since the formation of the Earth.

The total heat release includes the contribution of the descendants of uranium-238, including radium, which are in radioactive equilibrium with their ancestor and whose generally alpha decays lead to similar heat releases. The same applies to uranium-235 and thorium-232 families. On the whole, taking into account the decays of descendants amounts to multiply the energy release of uranium-238, uranium-235 and thorium-232 by around 10.

While the heat released by these radioactive atoms is small, the amounts of these radioactive elements in the Earth are also small. However, at the Earth scale, the quantities become enormous. The uranium and thorium amounts in the Earth crust and mantle are estimated respectively to 50,000 and 160,000 billion tons. According to this estimate, uranium alone would release the electrical energy produced by 4620 nuclear power plants of 1 Gigawatt.

The decay heat has somewhat decreased since the Earth formation, 4 billions years ago. Radioactive elements with a shorter lifespan that were present have vanished. The Earth enormous size guarantees that only a small fraction of this generated heat will emerge from the ground. The decay heat flux which is on the order of a few milliwatts per tonne for uranium and thorium, is lower for potassium because it contains very little radioactive potassium-40 (0.0117%), although Potassium is more abundant in the earth's crust.

Geothermal energy is one of the renewable energies considere today. Can one exploit this considerable source coming from the Earth, part of which is of radioactive origin? This energy is generally too dispersed to be recovered: The geothermal flux of 80 milliwatt per square meter would require an area of 1000 m^2 to power a single 80-watt bulb. On the other hand, heat accumulated in deep waters has been used for urban heating in the Paris Basin. Underground steam has been used to produce electricity in volcanic regions like Iceland.

Earth Crust Activity

By combining the volume of the earth crust with the density one can derive the average mass of the crust (depending on assumptions on crust thickness and density:

$$V_{shell} = V_{earth} - V_{innersphere} = \frac{4}{3}\pi \cdot \left(r_e^3 - r_i^3\right) = \frac{4}{3}\pi \cdot \left(r_e^3 - \left(r_e - d\right)^3\right)$$

$$V_{shell} = \frac{4}{3}\pi \cdot \left(6370^3 - \left(6370 - 40\right)^3\right) = 2.0 \cdot 10^{10}\,km^3 = 2 \cdot 10^{25}\,cm^3$$

$$\rho = 2 \cdot 5\frac{g}{cm^3} \Rightarrow M = \rho \cdot V = 5 \cdot 10^{25}\,g = 5 \cdot 10^{19}\,tons = 5 \cdot 10^{22}\,kg$$

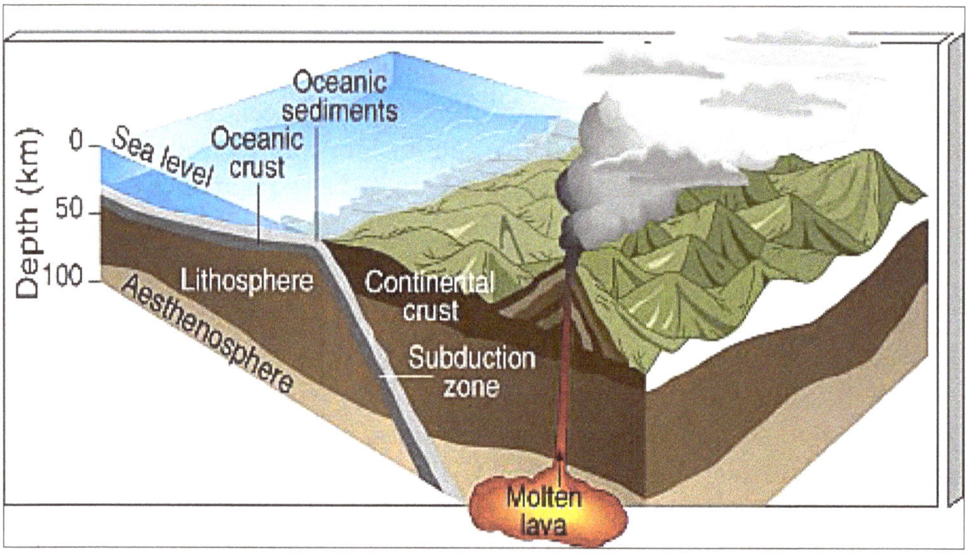

One can calculate the average radiation load distributed over the earth crust M(^{232}Th)=10 g/ton M(^{238}U)=2.5 g/ton M(^{40}K)=16800·0.00021 g/ton= 3.53 g/ton:

$$M\left({}^{232}Th\right)=10\frac{g}{tons}\cdot 5\cdot 10^{19}\,tons=5\cdot 10^{20}\,g,\ \ M\left({}^{238}U\right)=1.25\cdot 10^{20}\,g \text{ and } \ M\left({}^{40}K\right)=1.77\cdot 10^{20}\,g$$

$$N\left({}^{232}Th\right)=M\left({}^{232}Th\right)\cdot\frac{6.022.10^{23}}{232}=1.3\cdot 10^{42},\ \ N\left({}^{238}U\right)=3.2\cdot 10^{41} \text{ and } N\left({}^{40}K\right)=2.7\cdot 10^{42} \text{ particles}$$

Average Radiogenic Load

The number of radioactive nuclei in the earth crust allows us to calculate the average radiogenic load distributed over the earth crust. Reality shows pronounced distribution:

$$N\left({}^{232}Th\right)=1.3\cdot 10^{42} \text{particles},\ \ N\left({}^{238}U\right)=3.2\cdot 10^{41} \text{ particles and } N\left({}^{40}K\right)=2.7\cdot 10^{42} \text{ particles}$$

$$A\left({}^{232}Th\right)=\lambda\cdot N\left({}^{232}Th\right)=\frac{\ln 2}{T_{1/2}}\cdot N\left({}^{232}Th\right)=\frac{0.69}{1.405\cdot 10^{10}\cdot 3.14\cdot 10^{7}\,s}\cdot 3.1\cdot 10^{42}=2.03\cdot 10^{24}\,Bq$$

$$\Rightarrow\frac{A\left({}^{232}Th\right)}{M_{crust}}=\frac{2.03\cdot 10^{24}\,Bq}{5\cdot 10^{22}\,kg}=41\frac{Bq}{kg}$$

$$A\left({}^{238}U\right)=\lambda\cdot N\left({}^{238}U\right)=\frac{\ln 2}{T_{1/2}}\cdot N\left({}^{238}U\right)=\frac{0.69}{4.5\cdot 10^{9}\cdot 3.14\cdot 10^{7}\,s}\cdot 3.2\cdot 10^{41}=1.56\cdot 10^{24}\,Bq$$

$$\Rightarrow\frac{A\left({}^{238}U\right)}{M_{crust}}=11\frac{Bq}{kg}$$

$$A\left({}^{40}K\right)=\lambda\cdot N\left({}^{40}K\right)=\frac{\ln 2}{T_{1/2}}\cdot N\left({}^{40}K\right)=\frac{0.69}{1.28\cdot 10^{9}\cdot 3.14\cdot 10^{7}\,s}\cdot 2.7\cdot 10^{42}=4.64\cdot 10^{25}\,Bq$$

$$\Rightarrow\frac{A\left({}^{238}U\right)}{M_{crust}}=928\frac{Bq}{kg}$$

Be aware, these are average numbers due to initial assumptions and to large local fluctuations because of geological chemical and mineralogical processes.

References

- What-is-Geophysics, Education, Gobmx: sgm.gob.mx, Retrieved 17 August, 2020
- What-is-geodesy: gisresources.com, Retrieved 09 June, 2020
- Gravity-physics, science: britannica.com, Retrieved 11 May, 2020
- Gravimetry, earth-and-planetary-sciences: sciencedirect.com, Retrieved 13 August, 2020
- Radioactivity-lecture-251824: isnap.nd.edu, Retrieved 25 March, 2020

Sub-fields of Geophysics 2

- Environmental Geophysics
- Exploration Geophysics
- Borehole Geophysics
- Biogeophysics
- Near-surface Geophysics
- Tectonophysics
- Hydrogeophysics

Geophysics is branched into a number of sub-fields such as exploration geophysics, geodynamics, tectonophysics, borehole geophysics, hydrogeophysics, environmental geophysics, etc. This chapter closely examines these sub-fields of geophysics to provide an extensive understanding of the subject.

Environmental Geophysics

Environmental Geophysics deals with the estimation of physical properties in the shallow subsurface. The characterization of the near-surface structure is implemented by applying a variety of geophysical methods, i.e., active and passive methods. Active methods, such as seismics (reflection/refraction), electric/electromagnetic and radar techniques, use artificially generated fields on the earth's surface to further measure and interpret the earth's response. On the other hand, passive techniques such as magnetics, self potential and gravity, measure, analyze and interpret signals, supplied by the natural fields of the earth. The ultimate goal of the above mentioned geophysical techniques is to collect all the information regarding the near surface interior and to further contribute to the construction and protection of the human infrastructures, the protection of the life, health and heritage of living beings and the conservation of the natural resources. More specifically, environmental geophysics is involved in the investigation, detection and monitoring of:

- Construction works such as highways and railways, dams, bridges, abandoned pipelines and tanks, buried and unexploded ordnance-UXO and any related technical problem.
- Dangerous contaminants to public health and the environment such as LNAPLs, DNAPLs, heavy metals, fertilizers, in the soils and the groundwater.
- Geological hazards, such as sinkholes, landslides, karst collapse, faults and deformed stratigraphic units.
- Mapping of archaeological sites.

Environmental Geophysics generally applies non-destructive techniques to investigate the shallow subsurface in limited and wide areas, under certain conditions: 1. there are contrasts in the physical properties of the ground, 2. during the acquisition process, the optimal sampling parameters (in space and/or time) are selected, taking into account all available information corresponding to the ground conditions, 3. more than two geophysical techniques are applied in order to constrain the final solution and evaluate the results. 4. proper data processing, that comprises a crucial step in a geophysical project, as the selection of the wrong processing steps can lead to completely distorted results and further misinterpretation.

Robust Techniques of Environmental Geophysics

Well tested methods in Environmental Geophysics are seismic (surficial or crosshole passive or active seismic methods), electrical (1D, 2D, 3D or 4D electrical resistivity imaging techniques, time domain or spectral induced polarization methods, self potential) electromagnetic (including terrain conductivity, ground penetration, transient electromagnetic measurements), gravity-microgravity and magnetic (including magnetic susceptibility). In the present paragraph, the principles, the survey planning, the data acquisition, the interpretation and the limitations of the aforementioned methods are presented. Generally, when we plan a geophysical investigation, we should always bear in mind:

- To collect any information (geological, remote sensing, topographic, engineering, geophysical, etc.) concerning the study area.
- To know exactly the cost, the time of the survey and the size of the area.
- To estimate the possible noise sources in the local environment, affecting the applied geophysical methods, in order to decide the most effective combination. Noise tests should always be carried out prior to data acquisition.
- To estimate the detectability of the target and make statistical analyses concerning the expected range of the anomaly.

- To decide the expected lateral and vertical size of the target in order to select the appropriate sampling interval (ideally, at least 5 data points within the half width of the anomaly and the station/grid spacing).
- Forward modeling concerning the expected amplitude anomaly of the target.

Seismic Methods

Seismic Refraction Technique - SRefl

Seismic techniques can successfully reconstruct, with high resolution, the lateral and depth variation of seismic velocity, producing detailed images of the subsurface. The method is based on the fact that different geomaterials have different rock properties such as density, porosity, permeability and compaction, which can be recognized/recorded during seismic measurements. On the other hand, the amount of data usually collected during a routine fieldwork can rapidly become overwhelming and the logistics of data acquisition are more intense and expensive than in the other geophysical methods. In general, the equipment used for acquisition is more expensive than other instrumentation and data processing and interpretation can be time consuming.

In addition to the aforementioned limitations, there is one more inherent limitation related to the development of the method. The seismic method was originally developed by the oil industry and it was mainly intended to image the subsurface at depths of hundreds or thousands of meters. As a result, conventional seismic acquisition systems use very energetic sources and a large number of geophones. The use of seismic methods for imaging very shallow targets is a relatively recent development, and consequently its application in near surface experiments is still in progress. Shallow seismic experiments use portable seismic acquisition systems which have a limited number of channels, single geophones and relatively weak (and cheap) surface sources (usually a sledgehammer). In this case, a survey of this scale can be performed very quickly by at least two people, so that acquisition costs and time will be similar to those for a GPR survey.

To define the applicability of the seismic techniques in shallow application such as the environmental and engineering projects, one should consider which physical properties of the studied targets will make them detectable. In general, any velocity anomaly which is considered as a signal/target of interest can be defined in the shallower part (2-100m) of the earth. In some engineering applications, the depth of investigations can be higher depending on the structure and its foundation. In such applications, the seismic method can detect bedrock discontinuities, possible fracture zones, stratigraphy and water saturation which can be connected with liquefaction phenomena.

By collecting the data along a line, a seismic model was expected to be estimated applying refraction or reflection tomography. More recently, different kinds of data/information (seismic surface waves) acquisition have been developed to reconstruct the 2D/3D S-wave model of the study area. Seismic data can be collected into a borehole or a pair of boreholes deploying geophones and sources on the surface or into the boreholes, increasing the resolution of the resulted seismic model. Finally, depending on the size or the purpose of the project, the scale of the experiment can be changed, collecting and processing data in meso- and/or micro-scale.

The seismic refraction method can be applied for stratigraphic mapping, estimation of depth of the bedrock and the water table, prediction of the rippability of specific rock types, locating sinkholes, landfill investigations and geotechnical investigations.

Seismic Analysis of Surface Waves - SASW

Surface waves allow the measurement of the variation of S-waves in soils which can be transformed to stiffness with depth. Surface waves (Rayleigh and/or Love waves) have the property that ground motion becomes negligible below a depth of one wavelength. By recording surface waves of different frequencies (producing the dispersion curve), and therefore wavelengths, the properties of the subsurface can be determined at different depths. The characteristic velocity of the surface waves can be determined by measuring the signals received at a series of geophones. Once the shear wave velocity profiles are determined, shear (G_{max}) and Young's moduli of the materials can be calculated through the use of simple mathematical equations. These geotechnical parameters can be used by civil engineers for managing the foundation of large scale structures (bridges, dams, wind turbines, etc.). A stiffness profile can be generated for each adjacent pair of geophones. Profiles can be combined to produce cross-sectional images of the properties of the subsurface. Similarly, a number of adjacent cross-sections can be combined to produce a pseudo-3D image.

This method has been mainly used for engineering purposes, such as stratigraphic mapping prior any foundation, estimation of the depth of bedrock including landfill base location, soil stiffness and ground improvement verification and in-situ ground stiffness for large-scale geotechnical and civil engineering projects.

Seismic Reflection Technique - SRefl

The seismic reflection method has the highest resolution among the other seismic methods and identifies variations in material type with depth and horizontal position. The method was sponsored by the oil industry to provide detailed 2D and 3D images of the subsurface to the depth of thousands of meters. This technique images the interfaces between geological formations with contrasting acoustic or elastic or density properties and, consequently, contrasting seismic velocities. Mapping of these contrasts across an area can identify the extent and depth of specific layers or interfaces of interest (specific tectonic/stratigraphic structures which usually host oil called oil traps).

The SRefl method is based on the propagation of seismic waves through the subsurface and on their reflection at interfaces across, which there is contrast in velocity and/or other acoustic properties. In general, seismic energy is generated and the seismic waves travel through the subsurface. They are recorded by geophones at different positions along the survey line. Each geophone is connected to the seismograph (the acquisition system) which records the arrival time and magnitude of the induced voltages (oscillations) at each geophone.

Seismic reflection surveys are mainly used to map subsurface geological boundaries and stratigraphic variations. A key advantage of the technique is that, after detailed data processing, it can provide an accurate cross-section image of the subsurface.

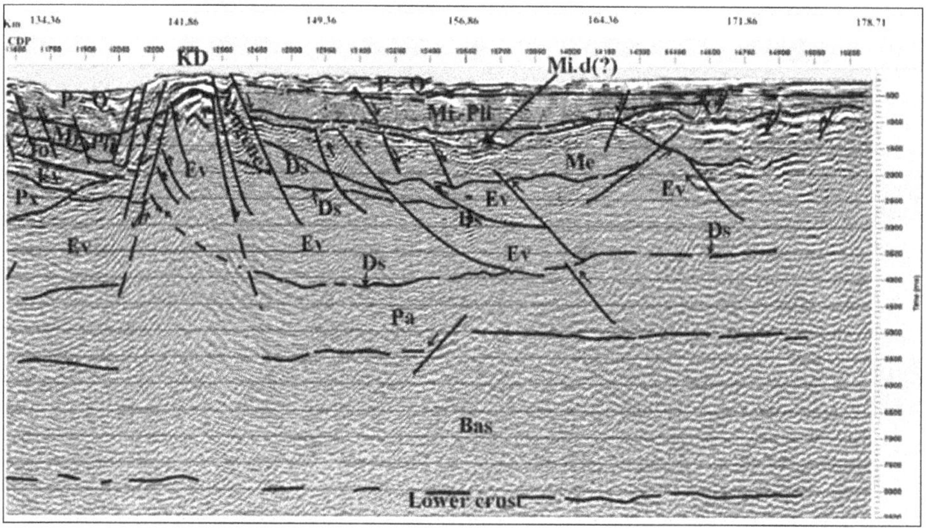

Interpreted deep seismic reflection section (migrated) from ION-7.

Advantages and Limitations of Seismic Methods

The conventional refraction seismic method has the following advantages and limitations: Advantages: Refraction measurements generally use fewer source and receiver pairs for producing a reliable subsurface model and thus, the method is relatively fast and cheap to be applied. A few processing steps are needed on raw seismic data to estimate the arrival times of the first breaks on recorded ground motion.

Disadvantages: Refraction only works if the velocity increases with depth (this problem was solved by using tomographic approaches solving the forward and inverse problem). The observations are generally interpreted in terms of layers-horizons. This method uses only the travel times between sources and receivers, excluding any other information that the waveforms have, such as the amplitude. Thus, the reconstructed velocity model is usually quite poor and has limited resolution.

The reflection seismic method has the following advantages and limitations. Advantages: Reflection measurements are collected at small source-receiver offsets to ensure near vertical ray paths. Processing and interpretation of the reflection seismic method are independent from the velocity variation with depth. Reflection seismics are able to reconstruct and depict complex geological models and make use of the whole waveforms.

Disadvantages: Since many source/receiver pairs must be used to produce detailed models of the subsurface, the method is very expensive. Reflection seismic processing can be very computer consuming, requiring huge CPU time for calculations and sophisticated software for processing by high-level expertise personnel. Thus, the processing of reflection seismic observations is relatively expensive.

The seismic analysis of surface waves has the following advantages and limitations. Advantages: The SASW/MASW (MASW - multichannel analysis of surface waves) technique is the optimum seismic technique for predicting the modulus of base materials in order to have a reliable geotechnical characterization of the study area. Due to multichannel recording (many acquired data) and processing schemes employed (many automated commercial software are available), results (Vs information) of the survey are highly accurate, even under the presence of higher modes of surface waves or cultural noise. For the same reason, the processing steps can be fully automated. Thus, the method is extremely userfriendly, easy and fast to implement.

Disadvantages: Due to the inner properties of surface waves, the resolution is highly dependent on the size of the anomaly. An empirical rule says that the minimum size that can be resolved is about one tenth of depth.

Electrical Methods

Electrical Resistivity Imaging – ERI

The electrical properties of the subsurface change due to the variation of the near surface earth material, with the presence or not of fluids that saturate the subsurface and to the occurrence of buried objects representing the target. Electrical techniques attempt to reconstruct the subsurface based on changes of the aforementioned properties distribution as a function of depth and horizontal distance. The most commonly used electrical technique is the Electrical Resistivity Imaging (or Electrical Resistivity Tomography, ERT) which can be applied in 2D, 3D and 4D (time lapse) mode. Simplifying the acquisition of ERI method, measurements of ground resistance are collected by introducing an electric current pulse into the subsurface using two metal electrodes (called current electrodes), planted onto the ground along a profile or into a borehole to collect well measurements. The current, passing through the subsurface, produces changes in electrical voltage (between two potential electrodes) in several places of the subsurface depending on the acquisition configuration. Applying the Ohm's law, the measured voltage can be "translated" to resistance reading for the subsurface between the two potential electrodes.

The ERI method can be used for landfill investigations, mapping and monitoring contaminated plume, groundwater pollution and infiltration studies, determination of the depth of the bedrock and location of sinkholes, buried channels, dykes, ore bodies, fissures, faults and mineshafts.

Induced Polarization Imaging – IPI

Induced polarization (IP) imaging is a complementary method to ERI and deals with the capacitance of the subsoil. The subsurface dissipates and stores the electric current flowing through it. The resistivity methods measure the resistance of the subsurface to energy flowing, whereas IPI method measures the amount of stored energy (referred to as chargeability). In theory, when a current flow through the subsoil, a small amount of energy is stored at the soil structure (minerals) and the soil becomes charged. When the current is turned off, this charge decays with time and can be recorded using potential electrodes. The estimation of the rate of this decay with time can provide the chargeability of the formation. Very often, IPI method can be very accurate since two formations can have similar resistivity but different charge abilities. This is the main reason that IP imaging is suggested as complementary method to ERI since it can provide additional information of the subsurface materials. The IPI method has the same applications as the ERI method.

Self Potential Method – SP

Self potentials (SP) are the difference in natural electrical potentials between two measuring points. These potentials can be produced by groundwater flow, contaminants movement, mineral deposits and chemical diffusion (electrokinetic phenomena). The magnitude (less than a millivolt to over one volt) and the polarity of self potentials are used for the interpretation of SP anomalies. The most common factor for SP anomalies is groundwater, and is generated by the flow of conductive water through the resistive medium, or by the involvement of water in natural chemical reactions. The SP method is mainly applied in hydrogeological investigations, landfill delineation, geothermal surveys, identification of ore bodies and detection of leaks in reservoirs and dams.

Advantages and Limitations of Electrical Methods

Advantages of the electrical methods are: The equipment is portable and inexpensive, compared with the instruments used in other geophysical methods. The user can have a qualitative interpretation of the collected data in a very short time quickly after the completion of the field measurement. Concerning the field crew, three experienced persons are enough to collect about 5-7 profiles per day depending on the total length of the profile and on the morphology of the study area. The method can be used for shallow or deep surveys, since the penetration depth depends on the electrode spacing and on the power of the instrument.

Disadvantages of the technique are: In cases of deep surveys (for mineral or geothermal explorations) larger electrodes spacing (about 20-50 meters) are required. Such long cables are heavy and consume much field time and personnel. The interpretation of the study area is not always straight forward since the geological and tectonic structure is complex. In such cases, a verification of the resulted tomographic images is needed using available borehole logs or other information which can constrain the resulted model.

Some of the advantages of IP data are: Having complementary IP data can help the interpreter to remove the ambiguity and reconstruct a better and more reliable subsurface model. The reason is that the IP measurements can distinguish the conductive clay from a conductive brackish saturated sand or gravel. The IP measurements are valuable to archaeological prospection and mineral exploration.

Gravity-microgravity

The gravity method was initially used in oil, gas and mining exploration and later in urban planning and engineering for detecting possible subsurface technical problems during construction works, in archaeology and in hydrologic investigations. A great deal of work has been done in the field of gravity, since it belongs to the oldest geophysical techniques. The gravity technique determines the earth's gravitational attraction, that is directly proportional to the mass and consequently to the density of the subsurface materials. By using the gravity method in geophysics, we are trying to specify density contrasts. A density contrast represents a gravity anomaly, corresponding to a local deviation of the background gravity field due to the presence of a subsurface geological feature or human construction (i.e., graves, tunnels etc.) with specific shape and geometry. The geological feature, causing the positive or negative gravity anomaly, could be a stratigraphic discontinuity or contact, a fault or a fold, a diapirism and an igneous intrusion. The units, commonly used in geophysics for the gravity anomaly are 1 mGal = 10^{-3}Gal, 1 gravity unit (gu) = 10^{6} ms^{-2} = 1 μms^{-2}, 1 gu = 0.1 mGal = 100 μGal.

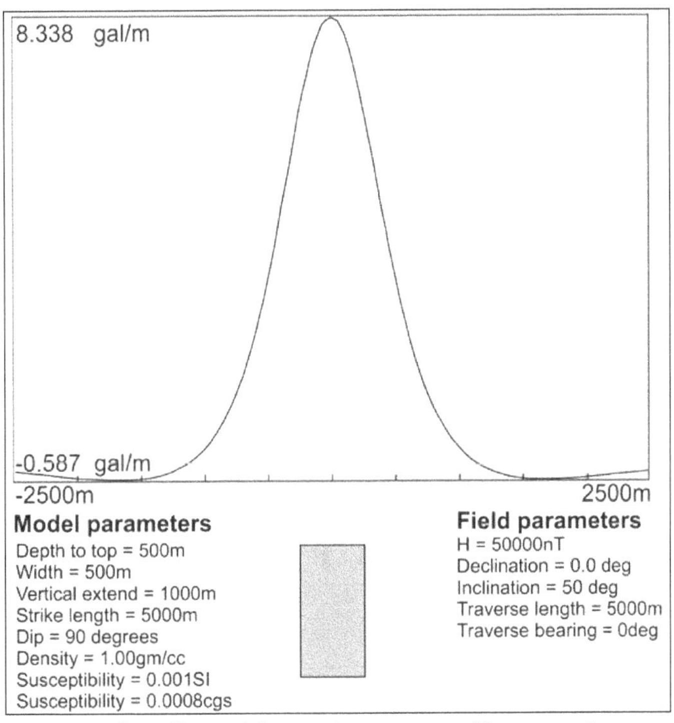

Vertical gradient of the gravity corresponding to a prism.

Advantages and Limitations of the Gravity-microgravity Method

The gravity-microgravity method is very effective in detecting targets that present a considerable density deference compared with the surrounding geological formations and targets underlain high velocity zones. Remote and difficult to access areas can be mapped using airborne gravity or marine gravity. The use of inexpensive workstations and personal computers has generally reduced the cost of this method (acquisition, processing and interpretation), while at the same time the quality and accuracy of the results has increased due to the rapid progress in software industry and the use of global positioning system (GPS).

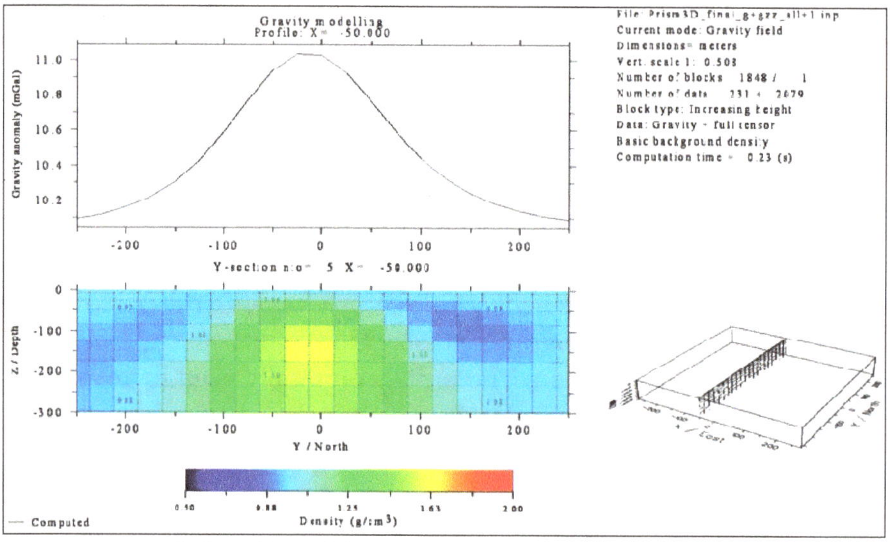

Vertical gradient of the gravity corresponding to a 3D prism.

The limitations of the gravity/microgravity method are:

- The gravimeter has to be placed very carefully on level ground.
- After every reading the instrument must be locked to avoid drift due to excessive vibration.
- The near and far topographic anomalies around the station, across the profile or in the grid, must be accurately mapped.
- Ground vibrations during the data acquisition strongly affect the accuracy of the value taken.
- The calibration of the instrument often needs more than one control points, that may be time consuming.

Magnetic Survey

Magnetic prospecting comprises a few centuries-old detection method, first applied in iron ore exploration. These days, the magnetic method is a fast and cheap method to acquire geophysical data for engineering and environmental purposes such as the detection of pipes, cables, tanks, drums, piles, metal objects and toxic waste, archaeological relics and old fire pits, geological features (magnetic basement, faults and folds) and finally for the analysis of tectonic processes. Magnetic measurements are used to estimate small, localized variations in the Earth's magnetic field.

In Environmental Geophysics, the magnetic unit used is the 1nT (1 billionth of a Tesla) or the 1gamma = 1nT. The magnetic response of a buried metal object may be a few hundreds of nT while archaeological relics (walls), in weakly magnetic sediments, correspond up to a few dozen of nT.

Although both magnetic and gravity methods belong to passive techniques, measuring the corresponding earth's potential field have some distinct differences between them: 1. The gravity field is due to one or more monopoles, while the magnetic field is always produced by a dipole, 2. The magnetic field is less stable than the gravity field, in terms that it is more sensitive to spatial and temporal changes.

Magnetic Susceptibility

Rock-magnetic parameters can be used for studying the environmental systems, the ground and marine soil magnetism, the fluvial processes, the lake sediments, the sources of magnetism in atmosphere, the palaeomagnetism and the biomagnetism. In particular, environmental studies based on magnetic properties attract the scientific interest in the last 30 years, focusing on the detection of contamination in sediments and soils in drainage systems, topsoils and vegetation, urban soils and ground pollution due to industrial emissions. The magnetic parameters of soils and rocks comprise a rapid and cheap indicator of industrially polluted areas and help in the targeted selection of samples for further geochemical analysis.

Magnetic susceptibility (X, as the ratio of the temporary magnetization to the applied field) is the most often used parameter in every environmental magnetism survey. More specifically, the magnetic susceptibility is a measure of how easy a material can be magnetized and is defined by the quantities of volume susceptibility k (dimensionless) and mass specific susceptibility X. Units of mass specific susceptibility of 10^{-6} m^3 kg^{-1} are commonly used in Environmental Geophysics. Accurate measurements of mass susceptibility are usually obtained in two frequencies (f_{low} = 0.43KHz and f_{high} = 4.3KHz). The magnetic behavior of many minerals changes with temperature. Curie temperature is the temperature point above which all magnetic minerals become paramagnetic. So, based on the shape of the temperature - susceptibility curves, the presence of magnetic minerals and domains can be detected. Anhysteretic remanent magnetization (ARM) and isothermal remanent magnetization (IRM) are often used in environmental magnetic investigations. Both properties are estimated by exposing a sample to an external magnetic field.

Advantages and Limitations of the Magnetic Method

The magnetic method is very effective in detecting low depth targets with high accuracy such as archaeological relics, geological features related to heat sources, intrusion of volcanic bodies, the aquifer and any kind of metal objects. The data acquisition and processing is fast and of low cost. Additionally, this method also works effectively in areas with irregular morphology. The limitations of the magnetic prospecting in Environmental Geophysics are:

- The survey area must be free of steel, iron metals and cables, because they influence the data acquisition. If an iron fence is present, measurements have to be taken more than 7-10m away.
- The magnetometer operator has to be free from any metal objects containing steel or iron. If the operator takes readings around the sensor and they do not change by more than 1 or 2 nT, data acquisition can start.
- The nature of the geological formation on ground comprises an important parameter affecting the measurements. For example "terra rossa," that is by its nature rich in ferrous minerals, generally causes noise during the data acquisition.
- Areas with dense vegetation and very irregular topography are generally more difficult to be surveyed.
- The magnetometer calibration is a very important procedure during a magnetic survey to improve the accuracy of target detection and localization. The instrument has to be calibrated following exactly the instructions of its manual, in the beginning of the survey and prior starting the measurements in the next grid, if many grids are to be surveyed. In addition, in case of sudden temperature changes the instrument has to be calibrated more often.
- Magnetic measurements must stop in case of strong magnetic disturbances of the earth's magnetic field.
- Deep targets are more difficult to detect and produce very weak magnetic responses.

The limitations of the magnetic susceptibility method in environmental geophysics are:

- Magnetic susceptibility comprises an indicator of possible heavy metals presence. The results of the pre-mentioned method should always be confirmed by geochemical analyses.
- Subsurface geological structure must be undisturbed in the surveyed area.
- The contact with metal objects has to be avoided during sampling.

- Magnetic susceptibility measurements on site must be taken away from metal objects.

Electromagnetic Methods

EM Terrain Conductivity (TCM) Method

In near surface electromagnetic (EM) surveys, the measured conductivity is presented as a function of depth along profiles or as a spatial variation. In general, the geological formations and the buried objects (geophysical targets) have different conductivities. By acquiring spatial (in X, Y) conductivity data anomalous areas can be identified. The TC method is based on the induction of electric currents into the ground by the magnetic component of electromagnetic waves generated at the surface. An alternating current, of variable frequency, is passed through a coil of wire (a transmitter coil). This process generates an alternating primary magnetic field which, in turn, induces very small eddy currents in the earth, the magnitude of which is directly proportional to the ground conductivity in the vicinity of the coil. These eddy currents then generate a secondary magnetic field, a part of which is intercepted by a receiver coil. The interaction between the primary and secondary magnetic flux and the receiver coil generates a voltage that is related to the electrical conductivity of the subsurface, expressed as milliSiemens/meter (mS/m).

Time Domain Electromagnetic Method - TDEM

The TDEM is a controlled source electromagnetic (CSEM) method for shallow (archaeological and groundwater) and deeper (geothermal, mineral, oil and gas) applications. Every CSEM system consists of a transmitter and a receiver, Tx and Rx dipoles (loops) respectively, with varying sizes depending on the required penetration depth. In general, an increase in transmitter loop increases the signal to noise ratio (S/N) resulting in an increase of the exploration depth.

In theory, a current flowing through the transmitter loop, generates a primary field. Applying an on-off switching sequence, a transient field is generated and is measured by the receiver loop. The decay rate of the electromagnetic field depends on the resistivity structure of the subsurface. Thus, the measured voltage on Rx coil can provide information about geoelectrical structures at several depths.

Ground Penetrating Radar

The Ground Penetrating Radar (GPR) method is a high-frequency electromagnetic technique commonly applied to solve engineering and environmental problems such as mapping of the shallow geological subsurface structure (stratigraphy and disturbances, water tables, voids and cavities), locating metallic and non metallic manmade structures (pipes, tanks, cables, landfill boundaries etc.) and characterizing archaeological sites.

The GPR technique partly works like the seismic reflection method, considering that source and receiver are located at the same point. The high frequency electromagnetic waves, produced by the source are transmitted in the Earth's interior and reflected on discontinuities dividing layers with different electrical properties (conductivity and the dielectric constant). The remaining part of the electromagnetic energy is diffused at deeper levels. The travel time of the electromagnetic wave from the source to a discontinuity and back to the receiver is recorded and if the propagation velocity of the electromagnetic wave is known then the depth of the reflector can be determined. Electromagnetic waves travel at the same speed in a vacuum/air, i.e., the speed of light $3x10^8$ ms^{-1}. Obviously, the speed of the electromagnetic waves travelling in the earth interior is much lower than $3x10^8$ ms^{-1}. The absorption of the electromagnetic waves is responsible for limiting the maximum penetration depth of a GPR. As this absorption is greater for higher frequencies, low frequency electromagnetic waves are able to propagate deeper in the earth. For instance, a GPR system working in the range of 25-50 MHz can investigate depths of over 50 meters in soils with low conductivity (smaller than 1mS/m) like sand and gravels. On the other hand, the spatial resolution of the method also decreases as the frequency is decreased.

In a profiling survey the GPR system consists of a transmitter and a receiver which are towed along transects. Radargrams are formed through the registration of the amplitude of the reflected signals and represent the time difference between the transmitted and reflected signals. Knowing the velocity of propagation of the EM waves over specific soils, it is possible to obtain a depth estimate of the suspected targets.

Advantages and Limitations of the Electromagnetic Methods

Terrain conductivity EM systems have the following advantages. For a daily survey, two people are needed. The first one carries a small Tx coil, while the second carries the Rx coil. TC meter is a very flexible and easy to use method, providing an average conductivity of the ground without any contact with the ground (as is required with DC resistivity techniques). The main disadvantage is the limited penetration depth.

The advantages of TDEM geoelectric sounding over conventional DC resistivity sounding (VES) are significant. They include the following: Surveys over large areas (from land, airborne, seaborne or into boreholes) with high lateral resolution. No contact with the ground is required. Improved resolution of conductive electrical equivalence and with little problem injecting current into a resistive surface layer. The limitations of TDEM techniques are listed as follows: It is not the optimum method for very resistive material. The penetration depth depends on used frequency and on Tx-Rx separation.

The advantages of the GPR system are: It can produce a detailed map of the vertical stratigraphy within a depth range of 1.5–10 m, depending on the operation frequency and the electrical properties of the ground. Features of sharp discontinuities and dielectric contrasts, such as walls, cellars, cavities, tombs and compacted earth can be quite easily resolved. On the one hand, radar can be easily used in urban areas; on the other hand, it exhibits serious problems in areas of conductive clayey soils because of the strong attenuation of the signal. Amplitude averaging techniques over specific depth ranges of radargrams obtained along parallel GPR transects provide depth (or time) slices of the subsurface and a three-dimensional (3D) reconstruction of it.

The limitations of the GPR in environmental geophysics are:

- The detectability of the instrument depends on the sensor's characteristics and on the target's dimensions/geometry.
- Metal is not penetrated by the radar waves and therefore objects behind a metallic boundary may be obscured and thus are not observable.
- Subsurface geology plays an important role in the depth penetration of the instrument. For example, sand comprises a formation easy to be penetrated by the electromagnetic waves while the clay not.
- It is usually recommended to survey almost flat areas, with no irregularities (no vegetation, rocks, holes etc.). However, it can also be effective in irregular ground, only the data acquisition is less efficient.

Data Acquisition, Processing and Interpretation

Forward and inverse modeling comprise the basic tools of data processing and further reliable interpretation in applied sciences, including Environmental Geophysics. Forward modeling is the procedure of synthetic data calculation, considering a known model of the earth's structure. On the other hand, the inversion is the procedure of estimating the Earth's model, often with constraints, using the instrumental data, provided by the survey.

Generally, the data processing flow of the geophysical data, and not only, is an iterative procedure and includes:

- Estimation of the error concerning the field observations, such as noise, location inaccuracies and any problems arising during data acquisition.
- Forward modeling prior the data acquisition, taking into account previous information, concerning the survey area. This step is crucial for the evaluation of the field data and to decide the flow diagram of data processing.
- Discretization of the earth's model, taking into account the previous information and the parameters of the data acquisition.
- Data fitting is the procedure showing how the predicted data (resulted from the forward modeling) matches the instrumental data. The well known least squares (L2 norm) and sometimes L1 norm (corresponding to the absolute differences between observed and calculated data) belong to the most wide used methods of data fitting.

- Inversion performance and evaluation of the results.
- Interpretation of the results.

Seismic Data Acquisition Processing and Interpretation

Seismic data (2D and 3D) can be acquired either on land or in marine environment. The field crew should have the ability to work on versatile field equipment, acquiring efficient, accurate, reliable and cost-effective 2D and 3D single and multi-component seismic data at any terrain and weather conditions. Of course, special attention should be taken to avoid erroneous measurements due to "strange" terrain and weather conditions. To gain time in processing and provide some preliminary results-models to clients, the field crew should offer a broad suite of in-field seismic data processing services. Based on the experience of the field managers, the exploration risk and cost can be reduced helping the clients receive satisfactory results sooner. Specifically, an accurate survey design can provide the highest quality and accuracy of data with minimum cost. Prior to any experiment in the field, the acquisition designers should estimate and provide the appropriate acquisition and parameterization parameters according to different field conditions to have the optimum imaging of the subsurface.

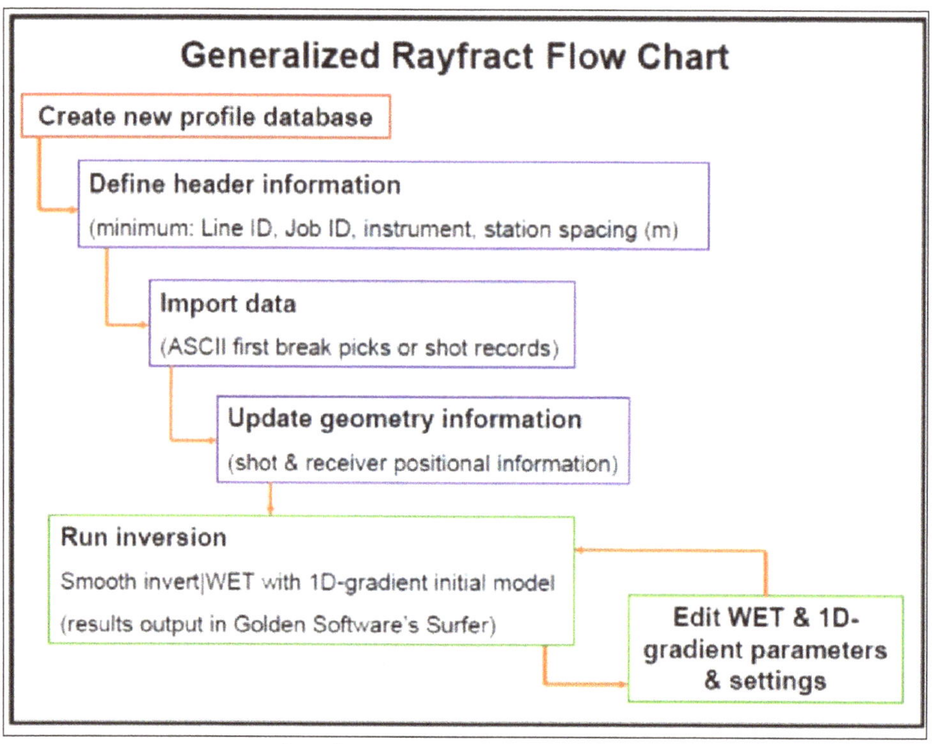

Flow chart of seismic data processing.

The operators should have specific knowledge and experience on, 2D and 3D (or even 4D) seismic survey design in different land conditions. VSP survey is designed especially for oil exploration and optimum hardware and equipment selection according to facilities and targets. The optimum equipment and acquisition parameters includes azimuth and offset distribution, geometry determination and the strategy of energy sources.

Seismic data processing is an important step before data interpretation and presentation of the final results. Usually, processing incorporates a wide variety of processing routines, such as: standard processing routines, velocity modeling, pre-stack depth/time migration and seismic imaging, ray tracing and tomography, multiples attenuation or elimination, seismic data reconstruction, data enhancement and multi-component data processing. Seismic data transcription, seismic data compression, seismic map digitizing and trace reconstruction are other complementary services.

Interpretation is the end of a very difficult way to knowledge. Months of preparation for fieldwork, data collection and processing are summarized in the final step of data interpretation. Having the appropriate knowledge, experience and the proper equipment to be efficient, enables us to perform seismic data interpretation. Since the interpretation needs a multidisciplinary approach, different kinds of available data like geological data, well logs, gravity and other geophysical data can be used to enhance the final model.

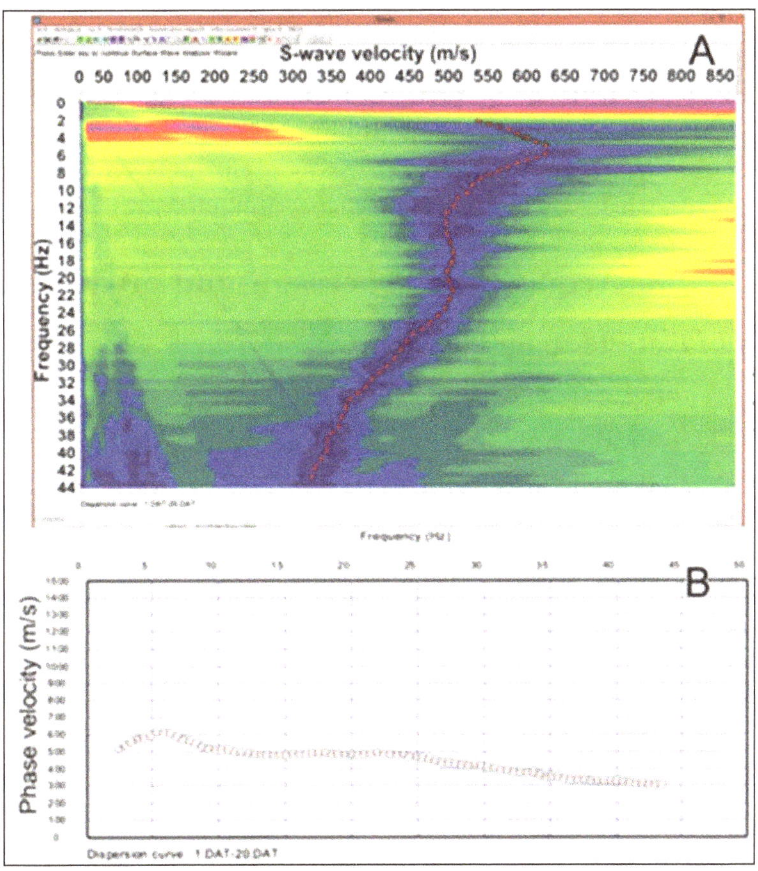

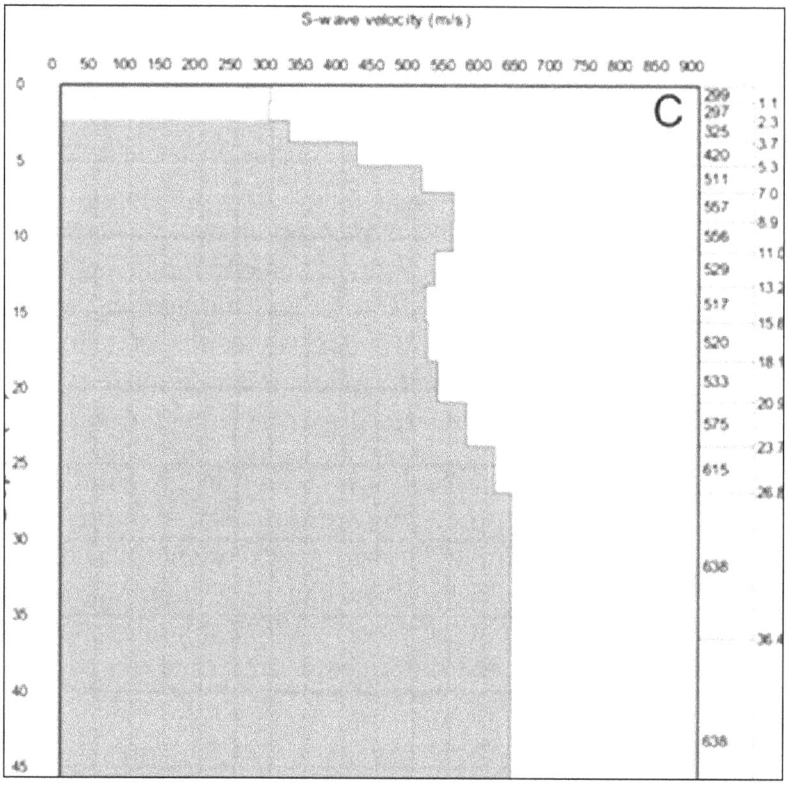

Processing steps of SASW data. A) Phase velocity-frequency plot (in color contours) is presented with the determined maximum amplitudes shown as red dashed line. B) The resulted dispersion curve is shown after processing of the phase velocity-frequency plot (A), C) Final Vs model after inversion of the dispersion curve as shown in (B).

Geoelectrical Data Acquisition, Processing and Interpretation

Acquisition of Geoelectrical Data

To construct an ERI cross-sectional image of ground resistance, a multi-electrode cable is connected with electrodes installed on the ground along a straight line (or a crooked line where the location of each electrode should be known to apply the appropriate corrections) with an inter-electrode spacing of 1-50 meters, depending on the requested depth of penetration. Once a measurement of ground resistance has been determined for a quadripole of electrodes, the next set of four electrodes is automatically selected and a second measurement of resistance is made. This procedure is fully automated and the time for the data collection is minimized due to modern instrumentation. This process is repeated until the end of the line is reached. The line is then re-surveyed with an inter-electrode spacing of 2, 3, 4, etc. Each increase in inter-electrode spacing increases the effective depth of the survey. The measured resistance values are converted to the value of apparent resistivity, (in ohmmetres) which can then be used to model the true subsurface resistivity distribution. The same acquisition system can be used for collecting IP data. Specifically, the measurements are made of both the resistivity and chargeability of the subsurface. Measurements of self potentials are made with two non-polarisable porous-pot electrodes connected to a high impedance voltmeter. Traditional metal stakes, as used in resistivity surveying, cannot be used as they generate their own potential when they are inserted into the ground. Data are collected along a survey line (SP profiling) or across a grid to produce a contour map of self potentials. The data require processing, as most interpretations are based on qualitative analysis of profile shape, polarity and amplitude. In geoelectrical methods, the orientation of the profile is crucial to ensure a high quality of data avoiding the anisotropy effects of the background geological formations.

Processing and Interpretation of Geoelectrical Data

Initially, data points, corresponding to wrong resistivity values, are removed. Next, filtering of raw data includes the removal of noisy points. After these preliminary steps, the user can continue processing by changing the inversion and regularization parameters or methods. Detailed instructions and information on how to process geoelectrical data can be found in several sites and manual of commercial software.

In many cases, the resulted tomographic models can be unreliable due to following reasons:

- Every reliablemodel needs good data: If the electrodes are not plugged well in the ground due to stones, hard soil or other reasons, or if the upper part of the investigation site is very dry (resistive) and the contact resistance is high or very conductive, the expected quality of raw data will be low. Try to solve such problems by changing the acquisition method/parameters or the geophysical method (applying seismics) in the field before pressing the start button.
- Non-uniqueness of inverse models: Same response can be resulted from different subsurface models. Geophysical methods and inversion can provide mathematical solutions with low error between observed and calculated parameters. Sometimes, depending on the complexity of the study area, this resulted model has nothing to do with reality. We should always use other kind of information such as borehole logs or other methods to constrain the model and construct a reliable geological model.
- Complex 3D geology: For processing, the most of available software assumes a 2D subsurface model. This assumption is true when the measurements are taken in proper orientations. If the geology is very complex, a distortion of the deeper sections of the resulted model is expected. For that reason, 3D processing software was recently used to give us an idea of the extent of the three-dimensional effects.

Gravity Data Acquisition, Processing and Interpretation

Gravity measurements are recorded along profiles, crossing the possible location of the subsurface target. The data are acquired using a base station and the mobile instruments. Sometimes for profile/grid of small length/area, and if the second instrument is not available, only one instrument can be used for both base station measurements and profile/grid measurements. The gravimeter, located in the base station, has to be on stable ground, concrete or not and protected from the weather conditions. The data of the base station are used to correct the grid data for instrumental drift and earth tides. For wide surveys, the base station measurement (the mean value of some readings to within 10 μGal) should be taken at the start and end of the day. At each point, in the profile/grid coordinates must be recorded.

Gravity data have to be corrected for elevation, influence of tides, latitude and, in the case of local topography irregularities, for topography. Bouguer gravity anomaly is the difference between observed gravity and theoretical gravity on the Earth's surface, after making all the required corrections. This is the final product used for the interpretation.

According to concerning the interpretation of a gravity anomaly:

- A positive gravity anomaly is due to feature of higher than average density.
- A negative gravity anomaly is due to feature of lower than average density.
- The dimensions of the target are proportional to the induced anomaly.
- A shallow feature induces a sharp high frequency anomaly.
- A deep feature induces a broad low frequency anomaly.
- The shape of the induced anomaly can be used to give more information on the depth, the shape and the total mass of the target.

Magnetic Data Acquisition, Processing and Interpretation

Magnetic measurements on ground are taken along a profile or in a grid at regular space interval, depending upon the target's depth and dimension. Shallow and small targets require surveys of higher resolution. The space interval is usually 0.5-1m and even smaller in very high resolution surveys. At each station, the mean value of generally 5-20 readings is taken. The magnetic response of a possible target is generally more complex compared with its gravity response due to the dipole nature of the magnetic field. This is clearly seen in figures, prepared by the same software, using the same field and model parameters, corresponding to a prism. However, in practice environmental magnetic data are often simpler to interpret than gravity because the noise and contributory sources are easier identified. The shape of a magnetic anomaly is often asymmetrical and varies depending on latitude and orientation of the causative body and the observed profile.

Magnetic susceptibility can be measured directly in the field or by collecting samples and placing them in plastic containers. Prior to starting the magnetic susceptibility measurements, the samples have to be mixed, air-dried, disaggregated and sieved, retaining the fraction smaller than 2mm in order to reduce the biasing effect of air, water and pebbles. Additionally, each sample is weighted and the susceptibility measurements in both frequencies are multiplied by a factor w_f = (10/weight of sample) in order to normalize the measurements for a mass of 10 gr.

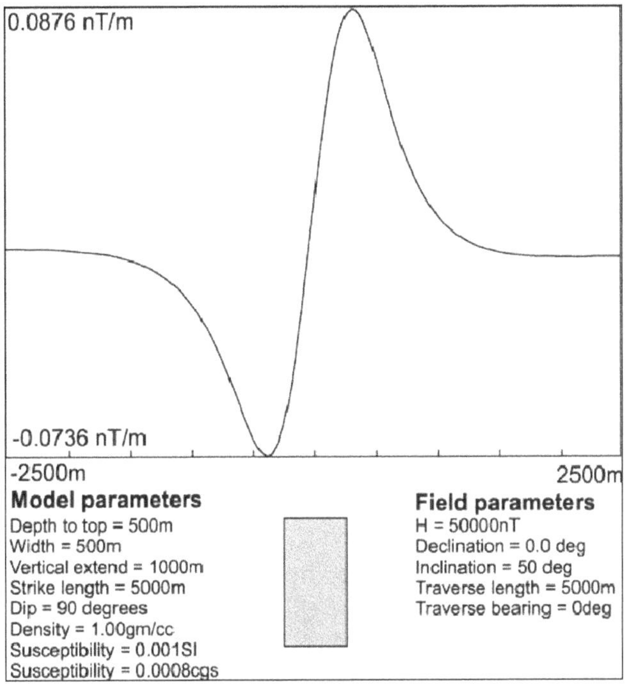

Vertical gradient of the magnetic field corresponding to a prism.

Generally, the magnetic data are initially corrected for the survey parameters and noise. Diurnal correction is applied in the field measurements for the temporal changes in the Earth's field during the survey, using the repeating measurements in a base station at frequent intervals. In the case of magnetic storms, magnetic surveys should stop. Normal field correction includes variations in field with latitude and longitude, in case of very large surveys. Elevation and terrain corrections are applied in any kind of magnetic survey (small or large areas). Finally, the regional field and the inclination-declination effects (reduction to the pole) are also removed from the data.

Magnetic data resulting from mapping small areas, such as magnetic survey of archaeological sites, are initially processed to correspond to a common base level (0-level base line) for all grids. This is to eliminate the shift of average value in the grids, due to differences in balancing the instrument and shifting of base/reference stations. The change of coordinates and the correction factors are used to create a synthetic mosaic of the grids. Maps are created through interpolation techniques. Selective de-spiking techniques and compression of the dynamic range of values are employed to isolate anomalies close to the background level. Grid equalization and line equalization are very useful techniques to smooth the data and to avoid stripping effects. Other filters such as high-pass filters (gradient) or the calculation of first horizontal derivatives are helpful in emphasizing the high frequency components of the magnetic maps.

Electromagnetic Data Acquisition, Processing and Interpretation

Terrain Conductivity Meters (TCM) data acquisition processing and interpretation. TCM read the apparent conductivity of the study areas in profile mode or acquiring measurements in grid (every 0.5 or 1m in X, Y direction). The distance between measurements is called the station spacing. In general, the station spacing should be between one-third to one-half the intercoil spacing. As soon as the target is found, additional geophysical methods are applied to reconstruct and refine the final subsurface model. Measurements in different coil orientations and/or intercoil separations can provide valuable information about the conductivity change with depth.

TDEM data acquisition processing and interpretation. The penetration depth depends on the size of the antenna (the bigger the loop size, the deeper the investigation), and on the type of the rocks. Three TEM measuring array configurations are used: central loop, offset loop and the coincident loop (or single-loop). The coincident loop configuration is the simpler, and perhaps the most usable. The same wire loop is used for both transmitter (TR) and receiver (REC), and major advantages of this loop configuration include good signal-to-noise ratios and a uniform response as a function of transmitter location. However, a homogeneous relief area is required. During the TEM fieldwork, the operator should be very careful for the best possible site selection, the correct installation of the loop, and other conditions which can produce noisy and bad quality data, like man-made noise (power and telephone lines, pipelines, fences, etc).

After the raw data collection, the data should be edited, filtered and smoothed prior than the final modeling and imaging. Editing is a very important part of the analysis of the data, since the operator can correct mistakes contained in the header of the field data (Name, TR, REC, coordinates), can exclude wrong points from the resistivity curves and can change the position of a point. Once the resistivity data are derived, the next stage is modeling of the data, as the ultimate goal is to deduce the subsurface resistivity distribution (resistivity-depth information) from these sounding curves.

In figure, an example of data smoothing is shown. Two more curves, except the resistivity curve, are given as result of the raw data smoothing. The first (upper) continuous curve approximates initial ρα(t) data and the second (lower) curve is the calculated dependence ρfull(t).

GPR data acquisition processing and interpretation. The basic signal processing of the GPR data includes: 1. Post processing gain, corresponding to amplitude corrections 2. DC bias removal, that is elimination of the data offset due to the electronic design, 3. De-banding through background removal, corresponding to removal of the signal mean value from every trace in the horizontal dimension, 4. Migration, in order to put the reflection points in their true location and 5. Hilbert transform corresponding to an estimation of the relative power within the signals.

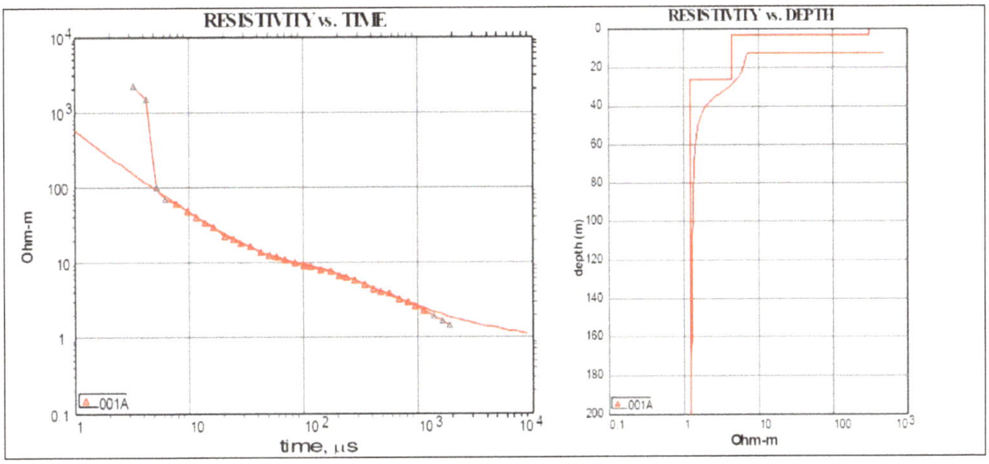

(Left) Curve of apparent resistivity ρa(t) versus time (example of sounding 001A).
(Right) The transformation (ρ(h)) and inversion (example of sounding 001A) resistivity model with depth are presented through the smooth curve and the piecewise-uniform diagram, respectively.

Exploration Geophysics

Exploration geophysics is an applied branch of geophysics and economic geology, which uses physical methods, such as seismic, gravitational, magnetic, electrical and electromagnetic at the surface of the Earth to measure the physical properties of the subsurface, along with the anomalies in those properties. It is most often used to detect or infer the presence and position of economically useful geological deposits, such as ore minerals; fossil fuels and other hydrocarbons; geothermal reservoirs; and groundwater reservoirs.

Exploration geophysics can be used to directly detect the target style of mineralization, via measuring its physical properties directly. For example, one may measure the density contrasts between the dense iron ore and the lighter silicate host rock, or one may measure the electrical conductivity contrast between conductive sulfide minerals and the resistive silicate host rock.

Geophysical Methods

The main techniques used are:

- Seismic tomography to locate earthquakes and assist in Seismology.
- Reflection seismology and seismic refraction to map the surface structure of a region.
- Geodesy and gravity techniques, including gravity gradiometry.
- Magnetic techniques, including aeromagnetic surveys to map magnetic anomalies.

- Electrical techniques, including electrical resistivity tomography and induced polarization.
- Electromagnetic methods, such as magnetotellurics, ground penetrating radar, transient/time-domain electromagnetics and SNMR.
- Borehole geophysics, also called well logging.
- Remote sensing techniques, including hyperspectral imaging.

Many other techniques, or methods of integration of the above techniques, have been developed and are currently used. However these are not as common due to cost-effectiveness, wide applicability and/or uncertainty in the results produced.

Uses

Exploration geophysics is also used to map the subsurface structure of a region, to elucidate the underlying structures, spatial distribution of rock units, and to detect structures such as faults, folds and intrusive rocks. This is an indirect method for assessing the likelihood of ore deposits or hydrocarbon accumulations. Methods devised for finding mineral or hydrocarbon deposits can also be used in other areas such as monitoring environmental impact, imaging subsurface archaeological sites, ground water investigations, subsurface salinity mapping, civil engineering site investigations and interplanetary imaging.

Mineral Exploration

Magnetometric surveys can be useful in defining magnetic anomalies which represent ore (direct detection), or in some cases gangue minerals associated with ore deposits (indirect or inferential detection). The most direct method of detection of ore via magnetism involves detecting iron ore mineralisation via mapping magnetic anomalies associated with banded iron formations which usually contain magnetite in some proportion. Skarn mineralisation, which often contains magnetite, can also be detected though the ore minerals themselves would be non-magnetic. Similarly, magnetite, hematite and often pyrrhotite are common minerals associated with hydrothermal alteration, and this alteration can be detected to provide an inference that some mineralising hydrothermal event has affected the rocks.

Gravity surveying can be used to detect dense bodies of rocks within host formations of less dense wall rocks. This can be used to directly detect Mississippi Valley Type ore deposits, IOCG ore deposits, iron ore deposits, skarn deposits and salt diapirs which can form oil and gas traps.

Electromagnetic (EM) surveys can be used to help detect a wide variety of mineral deposits, especially base metal sulphides via detection of conductivity anomalies which can be generated around sulphide bodies in the subsurface. EM surveys are also used in diamond exploration (where the kimberlite pipes tend to have lower resistance than enclosing rocks), graphite exploration, palaeochannel-hosted uranium deposits (which are associated with shallow aquifers, which often respond to EM surveys in conductive overburden). These are indirect inferential methods of detecting mineralisation, as the commodity being sought is not directly conductive, or not sufficiently conductive to be measurable. EM surveys are also used in unexploded ordnance, archaeological, and geotechnical investigations.

Regional EM surveys are conducted via airborne methods, using either fixed-wing aircraft or helicopter-borne EM rigs. Surface EM methods are based mostly on Transient EM methods using surface loops with a surface receiver, or a downhole tool lowered into a borehole which transects a body of mineralisation. These methods can map out sulphide bodies within the earth in 3 dimensions, and provide information to geologists to direct further exploratory drilling on known mineralisation. Surface loop surveys are rarely used for regional exploration, however in some cases such surveys can be used with success (e.g.; SQUID surveys for nickel ore bodies). Electric-resistance methods such as induced polarization methods can be useful for directly detecting sulfide bodies, coal and resistive rocks such as salt and carbonates.

Seismic methods can be also used for mineral exploration, since they can provide high-resolution images of geologic structures hosting mineral deposits. It is not just surface seismic surveys which are used, but also borehole seismic methods. All in all the usage of seismic methods for mineral exploration is steadily increasing.

Hydrocarbon Exploration

Seismic reflection and refraction techniques are the most widely used geophysical technique in hydrocarbon exploration. They are used to map the subsurface distribution of stratigraphy and its structure which can be used to delineate potential hydrocarbon accumulations, both stratigraphic and structural deposits or "traps". Well logging is another widely used technique as it provides necessary high resolution information about rock and fluid properties in a vertical section, although they are limited in areal extent. This limitation in areal extent is the reason why seismic reflection techniques are so popular; they provide a method for interpolating and extrapolating well log information over a much larger area. Gravity and magnetics are also used, with considerable frequency, in oil and gas exploration. These can be used to determine the geometry and depth of covered geological structures including uplifts, subsiding basins, faults, folds, igneous intrusions and salt diapirs due to their unique density and magnetic susceptibility signatures compared to the surrounding rocks, the latter is particularly useful for metallic ores. Remote sensing techniques, specifically hyperspectral imaging, have been used to detect hydrocarbon microseepages using the spectral signature of geochemically altered soils and vegetation.

Specifically at sea, two methods are used: marine seismic reflection and electromagnetic seabed logging (SBL). Marine Magnetotellurics (mMT) or marine Controlled Source

Electro-Magnetics (mCSEM) can provide pseudo-direct detection of hydrocarbons by detecting resistivity changes over geological traps (signaled by seismic survey), then minimizing the number of wellcats.

Civil Engineering

Ground Penetrating Radar

Ground penetrating radar is a non-invasive technique, and is used within civil construction and engineering for a variety of uses, including detection of utilities (buried water, gas, sewerage, electrical and telecommunication cables), mapping of soft soils and overburden for geotechnical characterization, and other similar uses.

Spectral-Analysis-of-Surface-Waves

The Spectral-Analysis-of-Surface-Waves (SASW) method is another non-invasive technique, which is widely used in practice to image the shear wave velocity profile of the soil. The SASW method relies on the dispersive nature of Raleigh waves in layered media, i.e., the wave-velocity depends on the load's frequency. A material profile, based on the SASW method, is thus obtained according to: a) constructing an experimental dispersion curve, by performing field experiments, each time using a different loading frequency, and measuring the surface wave-speed for each frequency; b) constructing a theoretical dispersion curve, by assuming a trial distribution for the material properties of a layered profile; c) varying the material properties of the layered profile, and repeating the previous step, until a match between the experimental dispersion curve, and the theoretical dispersion curve is attained. The SASW method renders a layered (one-dimensional) shear wave velocity profile for the soil.

Full Waveform Inversion

Full waveform inversion (FWI) methods are among the most recent techniques for geotechnical site characterization, and are still under continuous development. The method is fairly general, and is capable of imaging the arbitrarily heterogeneous compressional and shear wave velocity profiles of the soil.

Elastic waves are used to probe the site under investigation, by placing seismic vibrators on the ground surface. These waves propagate through the soil, and due to the heterogeneous geological structure of the site under investigation, multiple reflections and refractions occur. The response of the site to the seismic vibrator is measured by sensors (geophones), also placed on the ground surface. Two key-components are required for the profiling based on full-waveform inversion. These components are: a) a computer model for the simulation of elastic waves in semi-infinite domains; and b) an optimization framework, through which the computed response is matched to the measured response, via iteratively updating an initially assumed material distribution for the soil.

Other Techniques

Civil engineering can also use remote sensing information for topographical mapping, planning and environmental impact assessment. Airborne electromagnetic surveys are also used to characterize soft sediments in planning and engineering roads, dams and other structures. Magnetotellurics has proven useful for delineating groundwater reservoirs, mapping faults around areas where hazardous substances are stored (e.g. nuclear power stations and nuclear waste storage facilities), and earthquake precursor monitoring in areas with major structures such as hydro-electric dams subject to high levels of seismic activity.

BS 5930 is the standard used in the UK as a code of practice for site investigations.

Archaeology

Ground penetrating radar can be used to map buried artifacts, such as graves, mortuaries, wreck sites, and other shallowly buried archaeological sites. Ground magnetometric surveys can be used for detecting buried ferrous metals, useful in surveying shipwrecks, modern battlefields strewn with metal debris, and even subtle disturbances such as large-scale ancient ruins.

Sonar systems can be used to detect shipwrecks.

Forensics

Ground penetrating radar can be used to detect grave sites.

Unexploded Ordnance Detection

Magnetic and electromagnetic surveys can be used to locate unexploded ordnance.

Borehole Geophysics

Borehole geophysics is an aspect of geoscience that deals with the vertical analysis of the conditions of the subsurface and its fluid content. It is the technique employed in order to have a firsthand well evaluation in the exploration for hydrocarbons. One of the main results of Borehole Geophysical experiment is the Time to Depth relation curve. This curve can be used to generate an accurate calculation of interval, RMS and Average velocities. Those velocities helps us to relate our Seismic in Time to actual Geology in Depth. There are many other surveys that can be done in a Borehole other than the simple Vertical Seismic Profile (VSP), like Walk away, Walk above, 3D VSP and Walk around. Each one of those mentioned surveys gives different results and helps exploration in a different way.

Commonly used Borehole Surveys

Zero Offset VSP

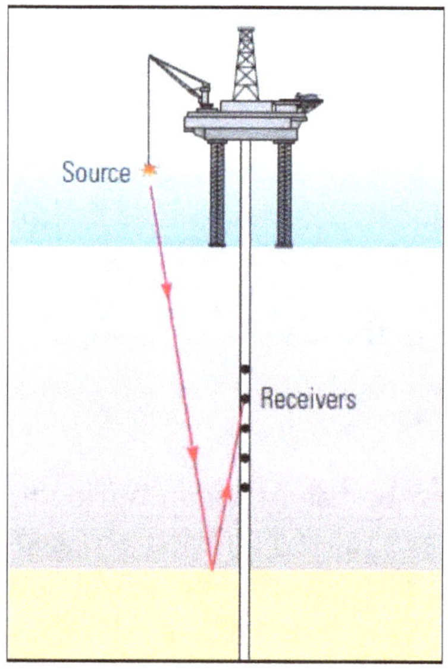

Its called zero offset because the source is very close to the wellbore, normally the source to well distance is around 200 m. This kind of survey normally provides the time to depth curve (check shots) and it can be used for multiples identification. As the source sends its waves downward, the receivers in the well bore receive those waves knowing the time of the shooting and the time of the first arrivals which connects our time to the tool Depth. The data will also show the formation reflectivity when the tool passes the formation top as a result of the reflectivity coming from the up going waves.

Deviated Well VSP/ Walk Above

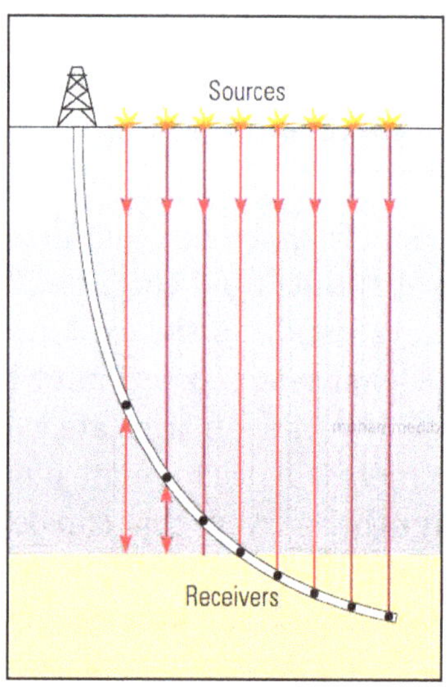

In Case of Deviated wells, the best approach is to shoot the source in a line that projects the well path from the top. The output of this survey is an image of the formations below the well path which is correlated to the surface seismic data except its higher in resolution.

Far Offset VSP

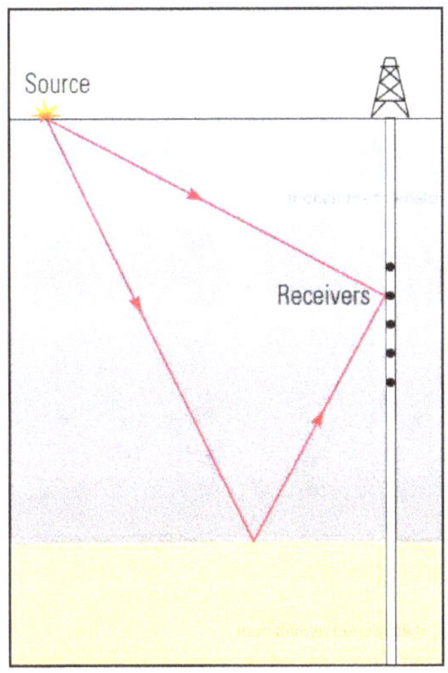

This kind of surveys is similar to the Zero offset, except the source is further away from the well bore. The offset increases the volume of subsurface imaged reflectors which increase the correlation with surface seismic data. Moreover, the offset provides higher chances of converting P waves into S wave. This provides a chance for amplitude variation with offset and anisotropy analysis.

Walk away VSP

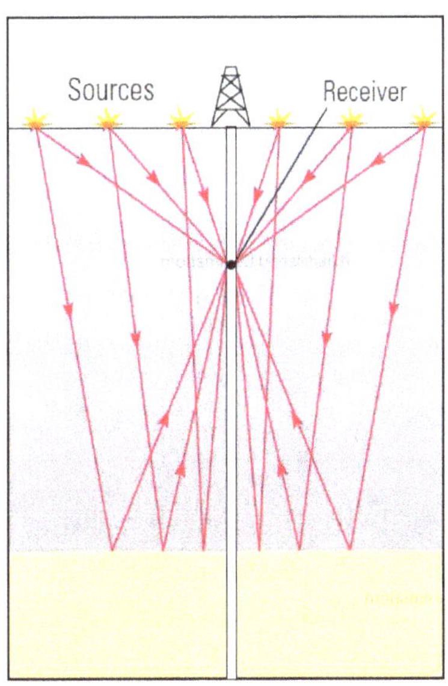

In the Walk away Survey, the receivers are fixed in the borehole while the source moves in a line that passes through the well head. Such survey is useful for shear Waves and anisotropy analysis. it can be compared to a 2D surface seismic except it includes both S and P waves with higher frequency bandwidth.

Walk around VSP

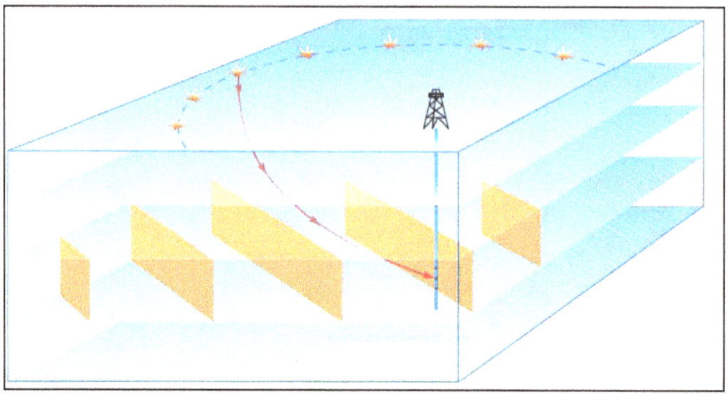

The Walk around survey is mainly specialized for anisotropy analysis, since the source circles around the well bore shooting, therefore any change of the received amplitude or frequency can reflect the pressure distribution in the area, assuming flat geology.

3D VSP

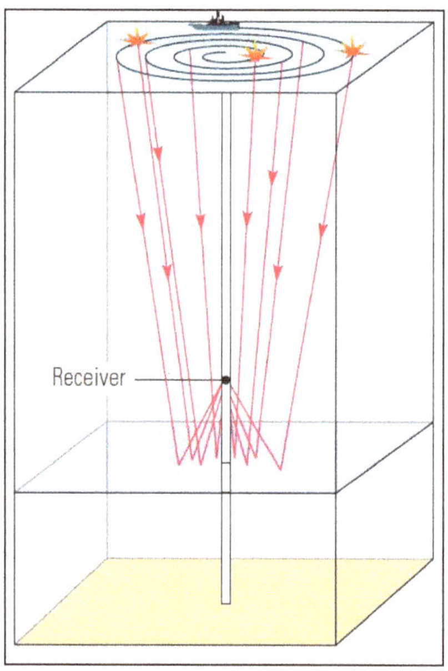

The 3D VSP survey is comparable to the surface seismic 3D survey except for the resolution difference, in addition to the areas that VSP fills in better than surface seismic in case of infrastructure difficulties, such as shallow gas which disrupts propagation of P waves. This kind of surveys is preformed in both land an marine, it can be also acquired in parallel with surface seismic 3D by putting receivers in both surface and in the wellbore.

Microseismic

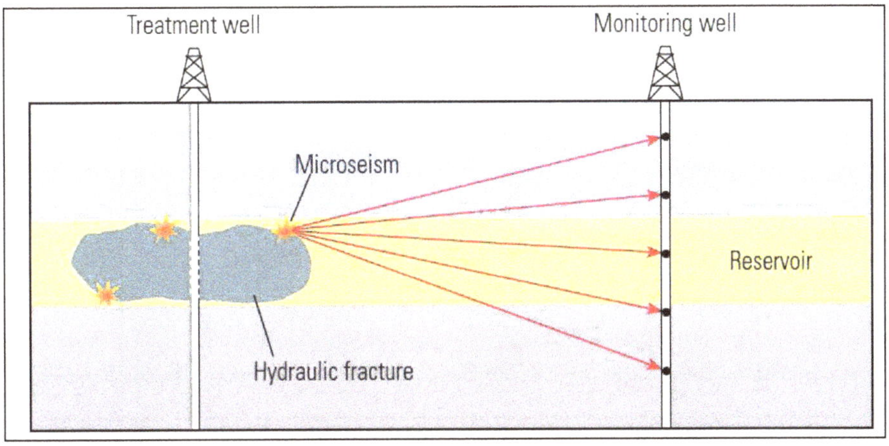

Another way is by monitory hydraulically induced fractures from nearby wells given an accurate velocity model to map the extent of the fracture. A different method is called passive seismic monitoring, which characterizes fractures by recording micro seismic signals generated when fluid is injected and produced modifying eh stress stated enough to cause seismic events. These micro seismic events can be plotted in space and time to identify the fractures responsible for stress changes. In some cases, the receivers are permanently installed withen the borehole due to the long periods of time needed for such surveys.

While Drilling VSP

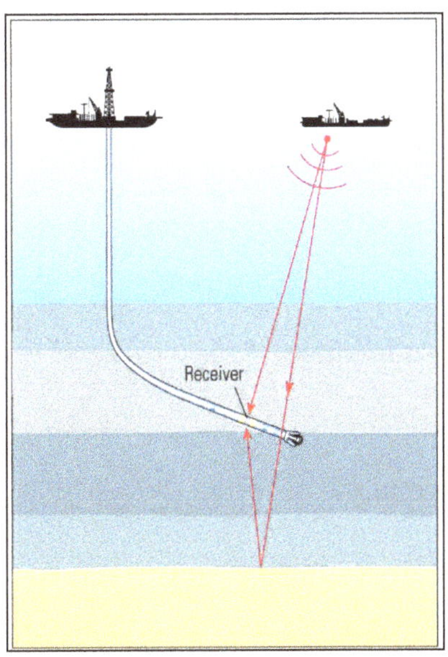

To help reduce uncertainty in time to depth correlation without having to stop drilling, geophysicist advice for seismic while drilling survey. This survey minimizes the risks that could be faced ahead of the bit, Such surveys can be done during pipe connection to avoid the noise generated by the bit while drilling.

Cross-well VSP

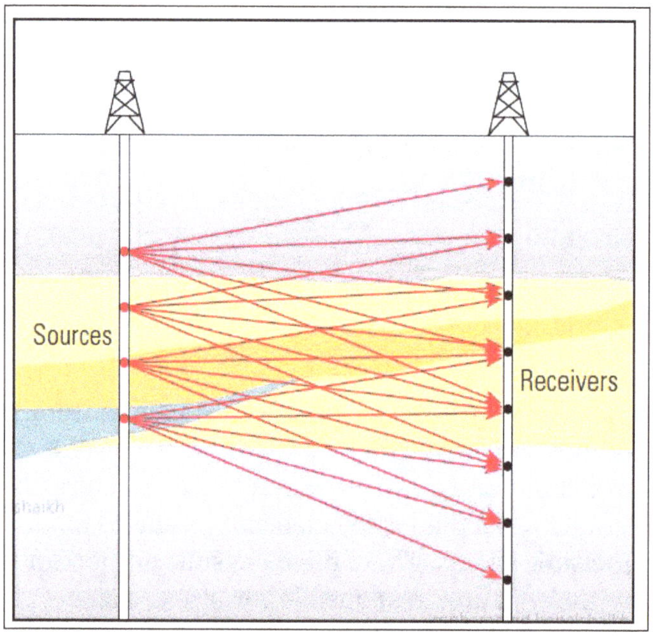

The source used in these kind of surveys are deployed at selected depth, while the receivers are placed in the nearby well. There is a limitation of the maximum offset to few thousand feets depending on the lithology at the area. Such surveys don't cover much of the reflectors as the rays travel parallel to them, however, a lot can be learned about the velocity local changes in the inter well region.

Types of Waves

- Primary P wave.
- Secondary shear wave.

Those waves can be found in 4 forms:

- Down going direct arrival.
- Reflected up going primary.
- Down going multiple.
- Reflected up going multiple.

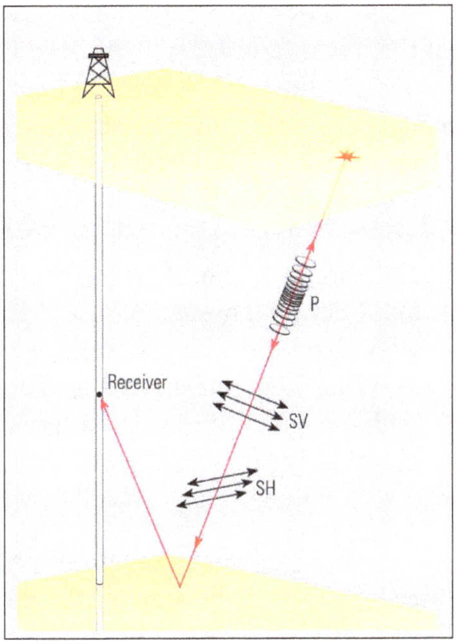

Biogeophysics

Biogeophysics is a sub-discipline of exploration geophysics focusing on the geophysical signatures resulting from microbial interactions with geologic media. "Biogeophysics" is defined as a rapidly evolving discipline of exploration geophysics concerned with the geophysical signatures of microbial interactions with geologic media that combines the fields of Microbiology, Biogeoscience, and Geophysics. Within this context, biogeophysics examines the links between dynamic subsurface microbial processes, microbial-induced alterations to geologic materials, and geophysical signatures.

Micro-organisms are ubiquitous and found in almost every conceivable niche of the Earth, from near-surface geologic materials as part of soil biota down to depths of tens of kilometers in the crust and the deep ocean, from hydrothermal vents to polar ice caps, and from alkaline to acidic environments. As such, micro-organisms have played an important role in many aspects of the evolution of the Earth and its atmosphere, leaving their mark on the geologic record. Their role in transforming Earth systems, the cycling of elements and formation of minerals, and the powering of Earth's biogeochemical cycles over the past 4 Gyr is well documented in many biogeochemical studies. From the initial stages of microbial colonization of mineral surfaces through the development of biofilms and biomats, the presence of micro-organisms within geologic media can have a profound effect on both the micro-organisms themselves as well as the chemical and physical properties of the surrounding environment.

Recognition of the important role that micro-organisms play in altering Earth systems resulted in the development of geomicrobiology (the study of microbial interactions with geologic media) as a scientific field of study more than 3 decades ago. However, their role in altering subsurface physical properties is only now being recognized and currently is not completely understood. "Although geophysicists have tended to carefully control the physical and chemical conditions during laboratory measurements of near surface materials, control of biological conditions, such as the level of bacterial activity has been completely overlooked." Yet it is apparent from decades of groundwater geochemistry studies that microbial metabolism and growth can impact both the chemical and physical properties of aquifers. For example, it has been documented from geochemistry studies that microbial metabolism can lead to the development of secondary porosity through the enhanced dissolution of different mineral phases or to the reduction of porosity by the clogging of pores due to precipitation of secondary mineral phases (e.g., calcite).

Furthermore, the growth of micro-organisms in the pore space, either in isolated colonies or as continuous biofilms, can lead to clogging of the porous medium with significant changes in the effective porosity and hydraulic conductivity causing the development of preferential flow paths. Porosity is an important aquifer and reservoir property and a key petrophysical parameter that can affect geoelectrical and seismic measurements. Hence, there is a growing interest in the geophysics community to investigate the links/relationships between biological interactions with geologic media and potential changes in geophysical properties in near-subsurface environments. Recognition of this potential has resulted in the development of a new subdiscipline in geophysics called "biogeophysics," which combines the fields of microbiology, biogeoscience, and geophysics. More specifically, biogeophysics is concerned with the links between dynamic subsurface microbial processes, microbe-induced alterations to geologic materials, and geophysical signatures.

Microbial Alteration of Petrophysical Properties

Our current understanding of how microbial communities and microbial interactions with subsurface geologic media leading to changes in petrophysical properties can be sensed with geophysical techniques. The main factors we identify are (1) the direct response from microbial cells and extracellular cellular structures, (2) growth of the micro-organisms (production of biomass) and their potential to form biofilms, (3) the generation of metabolic by-products and the interactions of these metabolites with the host material, and (4) microbe-mediated processes. Microbial growth and distribution in subsurface environments is dependent on many variables, including temperature, pH, physical and chemical properties of the geologic media (i.e., porosity and permeability), and availability of nutrients and electron donors and acceptors. Table gives the geophysical methods that have been used in support of biogeophysics research and provides what we consider to be the primary applications of each technique.

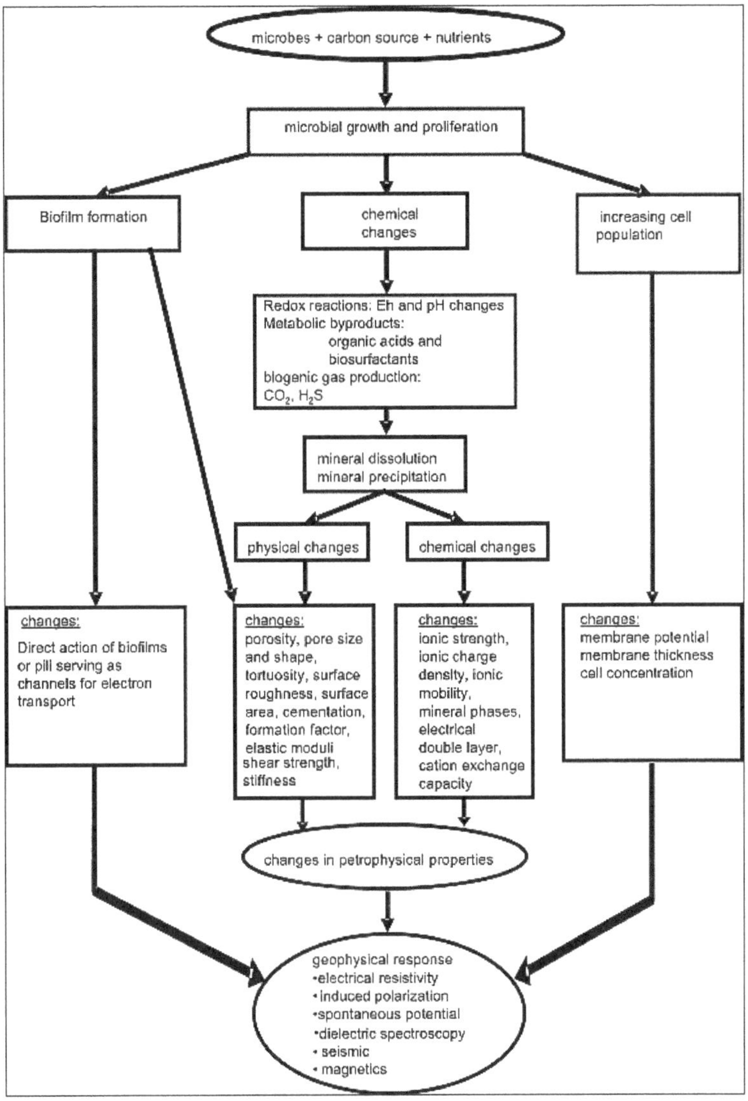

Biogeophysics flowchart illustrating the perceived relationships between microbiological/biogeochemical processes and geophysical responses.

Table: List of six geophysical methods that have been adopted for biogeophysics research.

Geophysical Method	Measured Properties	Measured Parameters of Interest	Biogeophysical Applications/Relevance
Electrical resistivity	Electrical potential generated by current injection	Dc conductivity (σ_{dc})	Biodegradation, biofilms?
Ground-penetrating radar	Amplitude and arrival time of electromagnetic energy (1 mhz to 1 ghz)	Dielectric constant (ε_r), dc conductivity (σ_{dc})	Biogenic gasses, biodegradation

Induced polarization/ spectral induced polarization/dielectric spectroscopy	Decay in electrical potential following current pulse, frequency dependence of complex impedance (mhz to khz)	Magnitude and relaxation length scale of dispersion caused by interfacial polarization	Biomineralization, cell properties (membrane potential), biofilms
Magnetometry	Magnitude and/or gradient of earth's magnetic field	Magnetic susceptibility	Magnetotatic batcteria, biomineralization?
Seismic	Amplitude and arrival time of elastic energy	Elastic wave velocities (V_p, V_s) and attenuation	Biomineralization, bioengineered soils, biogenic gasses
Spontaneous potential	Naturally occurring electrical potential, i.e., self-potential	Electrochemical potential, redox potential, streaming potential	Biogeobatteries, biodegradation

Microbial Cells and Biofilms

Although microbial cells can be free-living (mobile fraction), most exist attached to surfaces (immobile fraction). The advantages gained by the bacteria in living attached to a surface are thought to include (1) higher concentrations of nutrients close to a surface, (2) minerals sometimes serving as energy sources for microbial metabolism, and (3) protection and opportunities for synergistic relationships with other cells. The attachment of bacterial cells to mineral surfaces is influenced by the electrical properties of the bacterial cells which are related to the chemical composition of the cell wall. Because most bacterial walls (gram negative and gram positive bacteria) have higher concentrations of anionic groups, bacterial exhibit a net negative cell wall charge which is compensated by counterions resulting in electrical double layers (EDL) at their surfaces with high charge density (0.5–1.0 C/m^2). Hence bacteria are suggested to behave like electric capacitors capable of storing charges in the presence of an electric field. Static conductivity and dispersion studies show that the counterions give rise to considerable surface conductance within the hydrodynamically stagnant layer dependent on their concentration and the ionic mobilities of the counterions. Further, the interiors of living cells are electrically negative with respect to their environment. The finite membrane potential leads to an enormous low-frequency dielectric response of living cells at low frequency (10 Hz) in the presence of an electrical field but decreases at higher frequencies. Hence, low-frequency electrical techniques (e.g., dielectric spectroscopy or capacitance measurements) can be used to probe the electrical properties of living bacterial cells in suspension and in biofilm monitoring.

The bacterial EDL also plays an important role in adhesion/attachment of bacteria to mineral surfaces. There are several factors which potentially influence the adhesion of bacteria onto mineral surfaces, including bacterial species; the chemistry and surface area of the geological media; and the chemistry (ionic strength), pH, flow rate, and temperature of the pore water. Bacteria have a net negative charge on their surfaces, and therefore, as a result, bacteria experience strong electrical double layer repulsion on approach to negatively charged mineral surfaces. Conversely, the presence of metals at the surfaces of minerals (e.g., iron coatings) results in a greater affinity of bacterial adsorption to these surfaces due to the strong electrostatic attraction. During bacterial adhesion, a heterogeneous redistribution of charges on the bacterial cell surface reduces the EDL repulsion, enhancing attachment. Microbial adhesion to mineral surfaces has been the subject of numerous studies investigating microbial transport as this has direct implications for the fate and transport of contaminants.

Micro-organisms require both electron donors (usually a carbon source such as glucose, lactate, hydrocarbons, etc.) and electron acceptors (O_2, NO_3, Fe^3, Mn^4, and SO_4) for cellular growth and proliferation and to maintain metabolic functions. Biofilm formation requires the attachment of microbes to solid surfaces. Irreversible attachment occurs through the secretion of extracellular polymers that provide a strong bond to the solid surface that creates a matrix which surrounds the cells and forms the internal architecture of microbial communities. Further, during microbial growth, the increase in cell numbers and biomass (either as continuous biofilms or as isolated colonies) and associated extracellular polysaccharide structures can effectively clog/plug pores and change pore geometry and effective pore radii, leading to reductions in porosity, permeability, and hydraulic conductivity. In fact, pore clogging due to biomass production is a widespread phenomena documented in many natural and anthropogenic environments and engineered systems and is commonly observed during amendment strategies (i.e., biostimulation or bioaugmentation) for cleanup of contaminated sites.

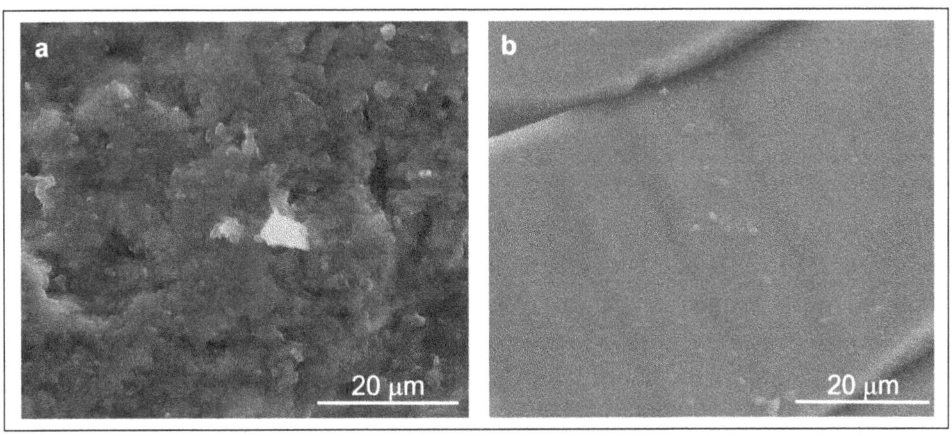

Environmental scanning electron microscope (ESEM) images of (a) cells of P. aeruginosaattached to a sand grain during a microbial growth experiment and (b) control.

The effect of bioclogging on hydraulic properties has been the focus of many experimental and numerical methods studies. Attachment of microbial cells to grain surfaces can also lead to an increase in surface roughness as well as surface area. For example, figure shows an environmental scanning electron microscope (ESEM) image of Pseudomonas aeruginosa cells embedded in biofilm and attached to the surface of a mineral grain acquired during a microbial growth experiment. Clearly, the surface characteristics of the mineral grain from the biostimulated column have been significantly altered compared to the grain from the unstimulated column. Both surface area and surface roughness are known to exert a major control on surface conductivity. Thus, the constriction of pores by biomass and the development of biofilms attached to mineral surfaces may have a profound effect on electrical properties, as we will discuss in section. Additionally, biofilms can act as organic precipitates and exhibit linear viscoelastic behavior. The presence of biofilms can fill pores and alter permeability and grain contact coupling thereby changing the sediment matrix stiffness, shear strength, and elastic and/or viscoelastic moduli of the media, potentially detectable by acoustic techniques and seismic techniques.

A recent discovery suggests that bacteria can produce certain proteinaceous extracellular appendages (pili) that are electrically conductive and serve to facilitate long-range electron transport. In addition to serving as electrical conduits for electron transfer to solid state electron acceptors (e.g., Fe(III) oxides), these pili (e.g., of Geobacter sulfurreducens) enable maximum biofilm growth also document that the pili of G. sulfurreducensenabled cells at the surface of a thick layer of biofilm to remain viable and connected to cells at the bottom of the biofilm in direct contact with an electrode, increasing electricity production by up to tenfold. Here we suggested that the pili of G. sulfurreducensprobably served as an electronic network permeating the biofilm and connecting cells throughout the biofilm, facilitating and promoting long-range electrical transfer in an energy-efficient manner. We speculate from the above study that in natural and anthropogenic environments in sufficient numbers, the permeation of these electrically conductive nanowires through porous media could conceivably increase the bulk electrical conductivity by providing a conductive network through an otherwise resistive geologic media. Conceivably, pili in biofilm across the water table interface may well represent a potential mechanism for electron transfer at organic-rich contaminated sites generating large self-potential anomalies akin to the geobattery effect of Sato and Mooney observed over sulfide deposits. Finally, we note that the action of magnetotactic bacteria can result in the precipitation of magnetosomes which can have a major impact on magnetic properties of the subsurface environment with a direct impact on magnetic geophysical techniques. Although in most field situations it is often assumed that the ferrimagnetic mineral content is too low to affect ground-penetrating radar (GPR) wave behavior, a study by Pettinelli et al. using magnetite/silica mixtures as Martian soil simulants observed that an increase in magnetite concentration affected both the wave velocity and the attenuation of electromagnetic (EM) waves. Therefore, the possibility exists that ferrimagnetic minerals produced by microbial activity in sufficient concentrations can have an effect on GPR signals.

Generation of Metabolic By-products

Microbial metabolism is considered to be a major factor enhancing the weathering of minerals through the attachment, growth, and colonization of mineral surfaces by micro-organisms. Microbial metabolic by-products include the production of biogenic gases (e.g., CO_2, H_2S, and CH_4), organic acids, and biosurfactants. These by-products have the potential to affect the physicochemical properties, and thus the electrical properties, of porous media in various ways. For example, microbial production of organic acids and biosurfactants (both cationic and anionic) directly adds ions to solution, increasing the electrolyte concentration of pore fluids, accounting for ~12% of fluid conductivity observed at organic-rich contaminated sites. In addition, evidence suggests that metabolic by-products such as organic acids enhance the mobility of sparingly soluble metals and also increase the rate of mineral surface reaction sites thus accelerating the rate of mineral weathering. The dissolution of minerals will further increase the ionic strength of pore fluids, enhancing the electrolytic conductivity (σ_w). In fact, in our own studies we have documented microbe-enhanced mineral weathering that resulted in direct increases in Ca^{2+} concentration, concomitant with an increase in the electrolytic conductivity and electrical conductivity. Enhanced mineral dissolution catalyzed by increased organic acid concentration can lead to physical changes in grain surface morphology, surface area, and surface roughness and the generation of secondary porosity and increased permeability. Changes in porosity resulting from mineral dissolution may alter grain contact coupling, affecting acoustic wave propagation. In addition, the generation of biogenic gases can have an effect on both the electrical conductivity (reducing conductivity) and seismic properties (enhancing attenuation of signal amplitudes).

Redox Processes

Microbial metabolic activity plays a key role in redox processes because microbes derive energy from oxidation-reduction reactions (the transfer of electrons from one reactant to the other). The chemistry of the fluid within the pores where microbial growth occurs is significantly different relative to the chemistry of the fluid in pores where microbial growth is absent. The abundance of terminal electron acceptors (TEAs) governs nutrient utilization by microbes during the breakdown of organic carbon. Terminal electron acceptors are sequentially utilized from O_2, NO_3^-, Fe(III), Mn(IV), SO_4^{2-}, and CO_2. As the TEAs are consumed, changes occur in the pore fluid chemistry. For example, products such as NH_4, Fe^{2+}, and Mn^{2+} may accumulate in the pore fluid. We have documented in field studies that peaks in the bulk electrical conductivity are coincident with peaks in redox species such as Fe^{2+} and Mn^{2+}. Significant changes in the pH and Eh in the sediment environment may also occur as a result of microbe-mediated redox reactions. Such pH and Eh changes can result in new mineral stability fields in which some minerals become unstable and are dissolved and mobilized (e.g., Fe^{2+}). On the other hand, as ions in solution exceed their saturation indices, minerals begin to precipitate from solution. For example, iron sulfide can be precipitated from solution during sulfate reduction by Disulfurbrio vulgaris. Therefore, micro-organisms and their activity can result in the formation of mineral phases by the process of biomineralization. Biomineralization can occur in two different ways: "biologically induced mineralization" in which the mineral precipitation occurs without control by the cell (passive mechanism) usually due to changes in microbial redox state (e.g., precipitation of calcite, magnetite, and iron hydroxides) and "biologically controlled biomineralization" in which mineral precipitation occurs by the micro-organism precipitating minerals that serve some physiological functions. The precipitates usually occur on mineral grains or in the pores between sediment. In some cases the bacteria cells can form the nucleation points of mineral precipitation, resulting in a biomineral. Mineral precipitation can result in changes in surface area, formation porosity, and pore throat diameter with significant impact on electric current flow and seismic wave propagation. Also, the precipitation of magnetite, for example, can change the magnetic properties of the subsurface contributing to the remanent magnetization of the rocks, while the weathering of iron mineral phases can equally contribute to the demagnetization of rocks.

The conceptual model presented in figure demonstrates that changes in the geophysical properties of geologic media are the end product of the integrated biogeochemical processes associated with microbial growth and activity. Microbial interactions with geologic media and their by-products can lead to changes in (1) the petrophysical properties of the rocks (e.g., changes in porosity, formation factor, tortuosity, and mechanical properties) by direct microbial colonization of mineral surfaces and biofilm formation or indirectly by mineral breakdown and precipitation from metabolic intermediates and end products, (2) the fluid chemistry by release of ions from mineral weathering and from metabolic intermediates (e.g., organic acids, biosurfactants, NH^{4+}, Fe(II), and Mn(II)), (3) the properties at the mineral surface-fluid interface by microbial colonization of the mineral surfaces or by concentrating chemical degradation products at this boundary, (4) the redox state of the groundwater geochemistry with the potential to develop strong electrochemical potentials, and (5) the geophysical signatures that arise directly from electrical and magnetic properties of cells themselves (e.g., precipitation of magnetosomes by magnetotactic bacteria). Figure also illustrates that microbial transformations of geologic media are complex and coupled and can occur over a wide range of spatial and temporal scales. Thus, the main challenge lies with decoupling the complex microbe-induced physical and chemical transformations inherent during microbial activity and quantifying the magnitudes of the geophysical response.

Direct Detection of Microbial Cells and Biofilms

Microbial cells themselves exhibit distinct electrical properties, and certain (magnetotactic) bacteria also display unique magnetic characteristics. Here we consider further the possibilities of detecting the geophysical signatures produced by bacterial cells themselves as well as those associated with biofilms forming in porous media.

Microbial Cells

One well-studied biophysical phenomenon is the polarization of living cells and the resulting macroscale dielectric permittivity of cellular suspensions. Laboratory biophysical studies of cellular suspensions have been primarily conducted with the focus on medical sciences, although more recent work has been conducted in support of national security, as well as fundamental research directed toward searching for evidence of life on other planets. Dielectric spectroscopy methods, analogous to electrical geophysical exploration methods, show that the low-frequency relative dielectric permittivity of suspensions of living cells is huge (in excess of 1000 below 100 Hz).

Broadband dielectric spectroscopy data on suspensions of live cells frequently record two polarization mechanisms, commonly referred to as the lower-frequency alpha (α) and higher-frequency beta (β) effect. The membrane potential (the voltage drop across the membrane due to the negatively charged molecules in cells) associated with live cells results in an accumulation of mobile electric charge carriers at the membrane surfaces. When live cells are placed in time-oscillating electric fields, these charges move on the surface of the membrane, giving rise to high polarizations (α effect). As the mobility of these surface charges is relatively small, this effect is manifest at low frequencies, such that the relative dielectric permittivity of live cell suspensions can be as high as 10^6. The effect is very similar to the polarization of the electrical double layer forming at the mineral surfaces of an interconnected porous medium, as traditionally measured with the spectral induced polarization (SIP) geophysical method. The β relaxation is related to the dielectric structure of the cell's membrane itself and can enhance the dielectric permittivity at higher frequencies.

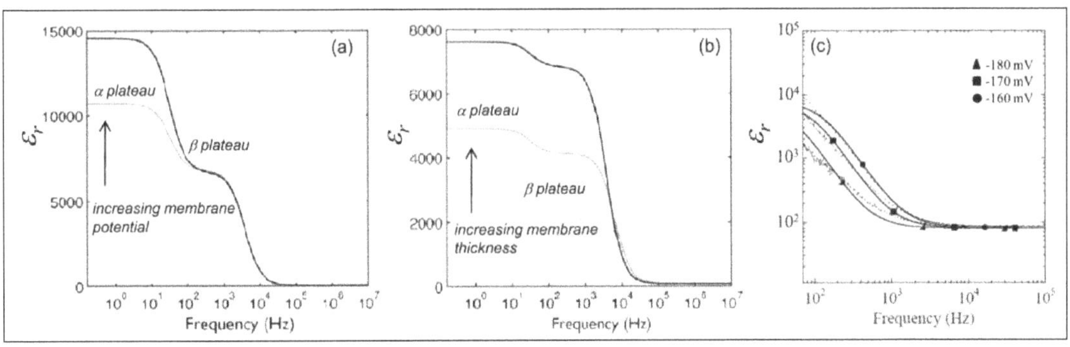

The effect of (a) cell membrane potential and (b) cell membrane thickness on dielectric spectroscopy data based on the analytical model of Prodan et al. The height of the αplateau is determined by the membrane potential whereas the height of the β plateau determines the thickness of the cell membrane.

Theoretical modeling frameworks have been developed to interpret dielectric spectroscopy measurements on cellular suspensions present a fully analytical solution for suspensions of spherical cells. The independent input parameters for this model are the outer and inner radius of the cell's membrane; the dielectric constant and conductivity of the medium, of the membrane, and of the inner cell region; the diffusion constants of surface charges accumulated at the outer and inner surfaces of the membrane; and the membrane potential. This model shows that the outer and inner cell radii, diffusion constants, and membrane potential all have a very distinct effect on the broadband dielectric spectroscopy data. For example, the membrane potential determines the height of the α plateau; the diffusion constants control the width of the plateau whereas the thickness of the cell membrane controls the height of the β plateau. Figure illustrates how the low-frequency polarization of Escherichia coli in suspension increases with membrane potential of the cells.

These well-known dielectric properties of cellular suspensions suggest that it may also be possible to use dielectric spectroscopy to investigate microbial cells in porous media. Ntarlagiannis et al. performed the equivalent of a single (10 Hz) frequency dielectric spectroscopy measurement on sand columns injected with varying high concentrations of Shewanella putrefaciens and E. coli and found measurable changes coinciding with changes in cell concentration inferred from optical density measurements. However, the interpretation of dielectric spectroscopy data in porous media is complicated by the additional possible polarization mechanisms that may exist relative to the two mechanisms (α and β) attributed to the electrical properties of cells themselves. For example, Ntarlagiannis et al. observed both increases (at higher cell concentration) and decreases (at lower cell concentration) in polarization resulting from the introduction of cells to the sand columns. They tentatively attributed the decreases at lower cell concentrations to alteration of the mineral-fluid interface due to cell-mineral interactions and the increases at higher concentrations to the development of a pore throat polarization mechanism resulting from clogging of pores by cells. Assuming this interpretation to be correct, they therefore did not detect the polarization due to the cells themselves, as seen with dielectric spectroscopy of cellular suspensions, but rather changes in the polarization of the porous medium resulting from the addition of cells in suspension.

More recent studies have shed additional light on the polarization mechanism resulting from addition of cells to a porous medium. Abdel Aal et al. conducted microbial adhesion studies to investigate the effect of cell sorption to mineral surfaces on induced polarization (IP) parameters. Clean quartz sands and varying percentages of iron oxide–coated sands were saturated with cell suspensions of Pseudomonas aeruginosa. They observed that microbial adsorption to clean quartz sands was gradual and resulted in an increase of 13% in the imaginary conductivity component (σ''). However, when iron oxide–coated sands (20%–100% by weight) were used, a more rapid increase in microbial adsorption was observed (due to the strong electrostatic force between oppositely charged surfaces) with σ'' reaching a maximum of 37% for the 80%–100% iron coated sands. In both experiments, no significant changes were observed in the real conductivity component (σ') due to adsorption of cells. A strong power law dependency was observed between the adsorbed cells and σ'', suggesting that the polarization (σ'') response resulted from the direct sorption of microbial cells to mineral surfaces. Abdel Aal et al. interpret these results to suggest that the increase in the surface roughness of the mineral grains due to the progressive sorption of bacteria is the main cause of the polarization and the large surface area of the bacteria accentuates this effect. Since most of the adsorption on cell surfaces took place within the first 90 min of the experiment, it is possible that part of the response observed by Ntarlagiannis et al. may have resulted from cell adhesion to mineral surfaces. Thus, the imaginary conductivity component appears a useful parameter for interrogating microbial cells and their interactions with mineral surfaces.

Magnetotactic Bacteria and Magnetic Signals

It is well known that a diverse group of prokaryotes can exert a significant control over magnetite formation. These micro-organisms, known as the magnetotactic bacteria, biomineralize intracellular, membrane-bound, single–magnetic domain crystals of the magnetic minerals magnetite and/or greigite called magnetosomes. In general, most magnetotactic bacteria can produce an average of 20 magnetosomes although some bacteria have been known to possess up to 1000 magnetosomes. The magnetosome crystals formed by magnetotactic bacteria have unique morphologies that are distinctly different from magnetite formed by Fe(II) reduction. These magnetosomes are synthesized from soluble forms of iron and are generally arranged as a chain that is anchored within the cell by cellular structural components. Using these magnetosome chains, like a compass needle, the cells are able to passively align along the Earth's geomagnetic field lines while they swim, aiding the cells to locate and maintain optimal position in vertical O_2 concentrations and redox gradients. Hence, magnetotactic bacteria are typically found in a wide variety of aquatic environments (e.g., freshwater lakes) with the highest numbers occurring mostly at oxic-anoxic interfaces (typically the sediment-water interface), indicating relatively specific redox and chemical conditions.

Their presence in sediments is often used as a proxy for paleoenvironmental/climatic conditions. Additionally, their presence in ancient sediments (magnetofossils) suggests that they are important remanence carriers and can significantly contribute to the environmental and paleomagnetic signals of sediments and rocks, making them useful for studying changes in the secular variation of the Earth's magnetic field from freshwater aquatic environments and even for the search for early life in ancient sedimentary formations. Studies estimate that the bacterial contribution to remanence in sediments is on the order of 10^{-6}–10^{-4} A/m. However, knowledge of magnetic properties of magnetosomes is still very limited, and their identification from bulk environmental samples remains a challenging task and an active area of research. While there have been no biogeophysical investigations per se that particularly address the issue of detecting magnetosomes using traditional geophysical techniques, we believe that the presence of magnetosomes in sediments and other secondary magnetic minerals produced by microbial activity can definitely impact the magnetic properties of the subsurface such that they may be measurable using magnetic geophysical techniques. For example, magnetic susceptibilitymeasurements at a hydrocarbon-contaminated site undergoing biodegradation showed a zone of enhanced susceptibility at depth interfaces coincident with the presence of contamination. The authors related the higher susceptibility values within the contaminated environment to the biotransformation of iron-related minerals within this zone.

Biofilms

A biofilm refers to an attached state of cell growth, whereby cells are closely packed and firmly attached to each other, usually forming a solid surface. Intuitively, the formation of biofilms might be expected to alter the geophysical properties of soils. Davis et al. conducted experiments on biostimulated sand-packed columns and found significant changes in column polarization coinciding in time with changes in concentration of cells attached to the sand surface. They used ESEM to show that peaks in polarization (at 2 Hz frequency) coincided with biofilm attachment to mineral surfaces and apparent bridging of soil grains by the biofilms. This led Davis et al. to suggest that low-frequency electrical measurements are a proxy indicator of microbial growth/cell attachment and biofilm formation in porous media. They attributed the initial increases in polarization to increasing cell attachment/aggregation into cell microcolonies. The later decreases in polarization were attributed to increased detachments rates or death and lysis of cells. Recent investigations by Abdel Aal et al. show that live microbial cells impact the magnitude of the phase and the imaginary conductivity components. Dead cells, on the other hand, have no impact on the phase or imaginary conductivity. These observations are consistent with the Davis et al. study.

Thus, the descending limb in the Davis et al. study is consistent with the death/lysis of cells. Biomass growth is self limiting because the growth of micro-organisms either as immobile fractions attached to mineral surfaces or as mobile components within the pores reduces the available pore space and nutrient flux, resulting in the eventual death of the micro-organisms. Hence, the Davis et al. study raises the interesting possibility that the imaginary conductivity component can be used to infer the rate of biofilm formation (growth) and decay (lysis) processes. However, it was unclear from Davis et al. whether the source of the polarization is the electrical properties of the biofilm itself (i.e., direct detection) or the modification of the polarization associated with the porous medium as a result of cell attachment to mineral surfaces (i.e., indirect detection). Ntarlagiannis and Ferguson attempted to further understand the low-frequency electrical properties of biofilms by conducting controlled experiments in sand columns injected with artificial biofilms created using an alginate gel matrix (alginate is known to simulate microbial exopolymeric substances. They showed that a low-frequency polarization developed in response to increasing volume concentration of (artificial) biofilm within columns packed with glass beads. They concluded that this experiment, unlike the experiment that Davis et al. conducted on a soil undergoing biodegradation, confirmed that a measurable polarization results from the presence of biofilm itself in a porous medium. Ntarlagiannis and Ferguson attributed this polarization to a pore throat mechanism generated as biofilms filled and blocked pore throats. However, the generation of such a polarization mechanism due to pore blockage by natural biofilms remains unproven but feasible since numerous studies have documented that the growth of microbial biomass in the form of biofilms in pores plugs the pore space, altering flow paths through porous media and reducing hydraulic conductivity and mass transport properties.

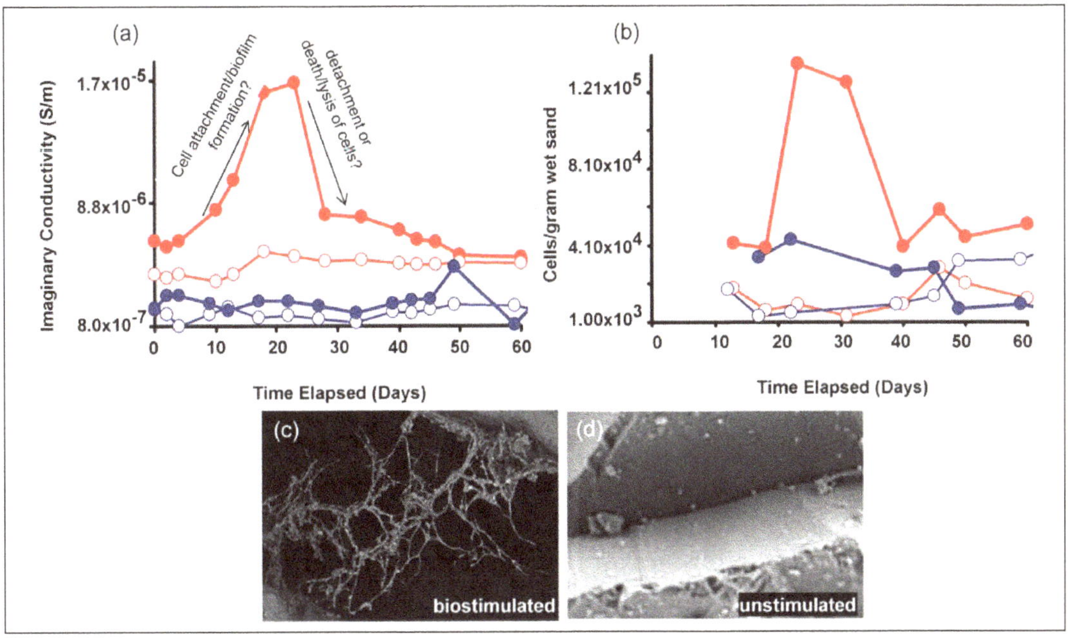

(a) Polarization signal associated with (b) density of cells attached to mineral surface as observed by Davis et al. Filled red circles are biostimulated column, and other symbols are control columns. Also shown are ESEM images from (c) biostimulated column showing cell attachment/biofilms and (d) unstimulated column.

The utility of using acoustic imaging techniques to investigate biofilm growth has recently been demonstrated by Davis et al. In controlled column experiments. In this study, microbial growth (Pseudomonas aeruginosa PAO1 wild-type bacteria culture) was stimulated in silica sand-packed columns, and acoustic wave (compressional wave) data over a two-dimensional region were collected for 29 days. The results showed relatively high spatial variability in the acoustic amplitude (with areas of enhanced and attenuated signals) measured from initial conditions of homogeneous amplitudes in the biostimulated columns while minimal changes were observed in the unstimulated column. The bulk of the changes to the acoustic wave properties in the biostimulated column were imparted between days 5 and 7 of the experiment, concomitant with a peak in the imaginary conductivity values. Davis et al. had previously demonstrated that peaks in the imaginary conductivity correlate with peaks in biomass production.

Hence, the commonality of when maximum changes occurred in the electrical and seismic measurements led the authors to suggest that the changes in the acoustic wave amplitude resulted from the direct presence of biomass accumulation. The spatial variability observed in the acoustic amplitude results from this study suggests that enhanced microbial growth has a variable affect on the elastic properties of porous media. While the exact microbe-induced mechanisms for the variations in amplitude are yet unclear, the authors speculated that the differences in amplitude may arise from a nonuniform distribution of microbial activity or possible heterogeneity in the biomass distribution and biofilm morphology (e.g., variations in biofilm thickness, roughness, and hydration). ESEM imaging qualitatively showed apparent differences in the structure and/or thickness of biofilm on the surface of the sand grains, as well as evidence of bacterial attachment, between areas of increased and decreased acoustic wave amplitude. Although significant changes in attenuation were observed, only slight changes in velocity were observed within the biostimulated columns.

The small change in velocity observed in the Davis et al. study these results suggest that microbial processes cause significant changes in the amplitude response of acoustic waves but only minimal changes in velocity. Geophysical techniques are sensitive to the growth of microbes in porous media and represent a significant first step toward the use of geophysical imaging techniques for the validation of bioclogging models, while highlighting the potential for using geophysical methodologies to provide spatiotemporal changes in biofilm growth and accumulation within porous media at field sites as well as mapping preferential flow paths resulting from this process. We also see some potential applications of geophysics for the assessment of biofilm barriers used for contaminant remediation or as sealants of reservoirs for CO_2 sequestration and the remote monitoring of microbial activity during microbe-enhanced oil recovery.

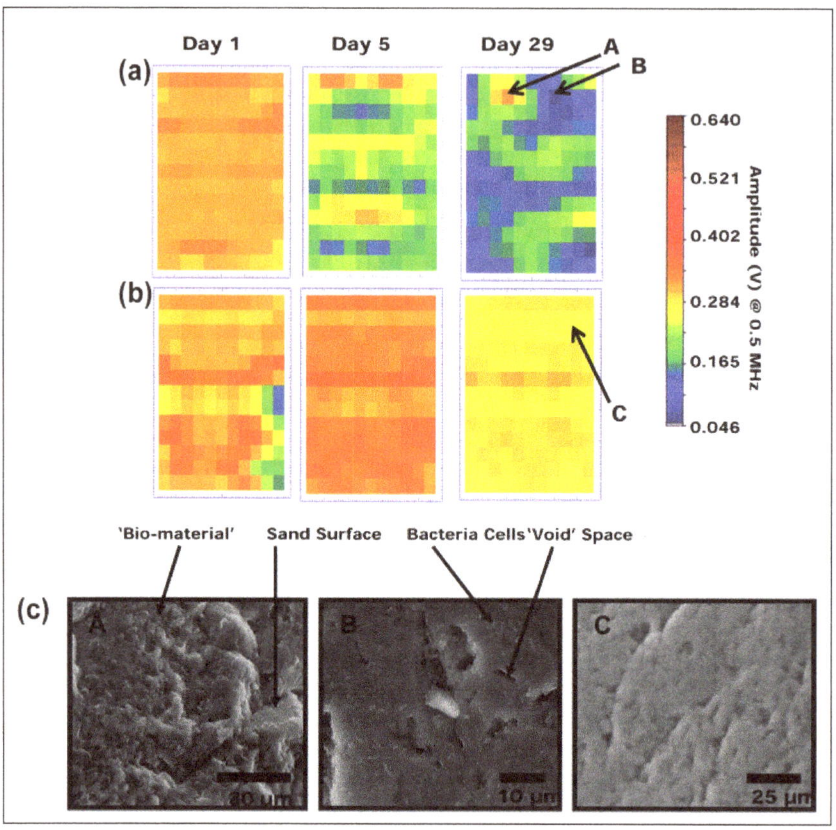

Two-dimensional seismic scan image from a biofilm growth experiment at selected times over a 29 day experiment: (a) biostimulated sample, (b) control sample, and (c)

ESEM images from sample locations shown in figures (a) and (b). ESEM images A, B, and C are from locations A, B, and C shown on the seismic scans

One of the most profound effects of microbial metabolism on the rock record is in the enhancement of mineral weathering either through mineral dissolution or mineral oxidation. The physical attachment and colonization of indigenous micro-organisms to mineral substrates in their attempt to gain access to nutrients promotes chemical weathering at rates faster than is theoretically predicted. In addition, the production of metabolic by-products such as organic acids enhances the dissolution of minerals, leading to significant increases in the total reactive surface area of minerals. Thus, micro-organisms play a fundamental role in the modification of rates of mineral weathering with significant implications for both soil and sediment formation and in the regulation of global climate as a result of their participation in the cycling of carbon through the different carbon reservoirs.

Hydrocarbon-contaminated environments represent one environment where the effect of metabolic by-products on geophysical properties is well documented. Until recently, it was commonly assumed by many field geophysicists that hydrocarbon-impacted sediments can only be effectively imaged by their higher resistivity (low conductivity) compared to "background" due to a simplified conceptual model based on the partial replacement of conductive groundwater by highly resistive petroleum compounds. However, extensive evidence exists in the literature for enhanced mineral weathering in hydrocarbon contaminated aquifers. This weathering results primarily from the production of organic acids and CO_2 (carbonic acid). For example during the congruent dissolution of calcite, Ca^{2+} and HCO_3^- are produced that go into solution:

$$CaCO_3 + CO_2 + H_2O \leftrightarrow Ca^{2+} + 2HCO_3^-$$

The net effect is the increase in the concentration of dissolved solutes that results in an increase in the pore fluid conductivity. We also note that elevated pore fluid conductivity can result from the direct addition of organic acids and biosurfactants to solution. Hence, elevated pore fluid conductivity at hydrocarbon contaminated sites appears to be the rule rather than the exception. Changes in pore fluid conductivity directly result in changes in bulk conductivity measurable by electrical resistivity, SIP, electromagnetic induction, GPR, and self-potential (diffusion potentials) techniques. Indeed, the effect of metabolic by-products such as organic acid production on electrical properties is well demonstrated by the seminal work of Sauck et al. at the fire training facility (FT-02) at the decommissioned Wurtsmith Air Force Base, Oscoda, Michigan, United States. The electrical resistivity method and GPR investigations across the hydrocarbon plume at this site showed elevated bulk electrical conductivity and attenuated GPR reflections coincident with the zone of contamination. Groundwater investigations suggest that intrinsic bioremediation and enhanced mineral weathering are occurring at the site.

In addition, studies by Bradford at the same site observed a distinct zone of increased dispersion coincident with the zone of GPR signal attenuation. The increase in the pore water conductivity alters the relaxation characteristics of the bulk formation, resulting in the attenuation of the GPR signal amplitudes. Attenuated GPR reflections resulting from elevated pore fluid conductivity due to biodegradation have also been reported by Cassidy. Using GPR signal attribute analysis and dielectric property studies, Cassidy documented attenuation related to material property losses of >2–6 dB/m within contaminated zones. The author also noted a slight but observable frequency dependence to the attenuation spectrum with the lowest-frequency data (<300 MHz) showing the greatest degree of loss.

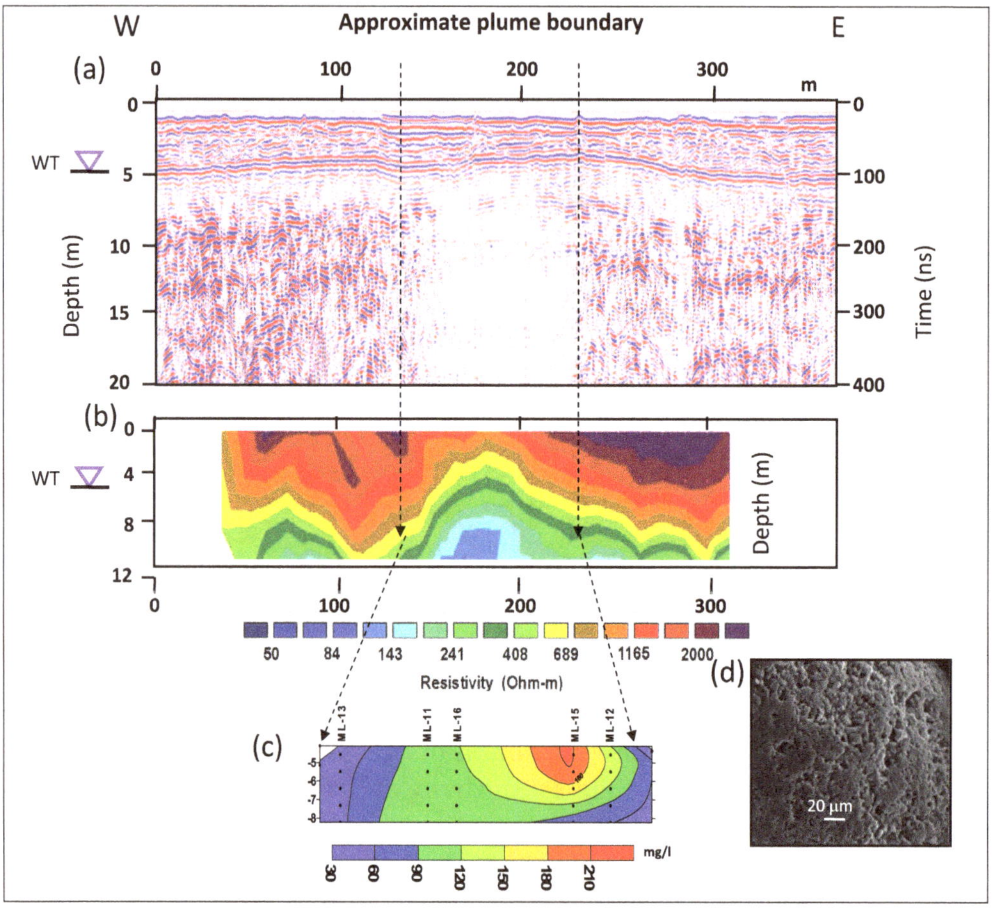

(a) Ground-penetrating radar profile across a hydrocarbon contaminated site showing a zone of attenuated reflections coincident with the plume, (b) electrical resistivity across the plume (red indicates high resistivity, and blue indicates low resistivity), (c) calcium ion concentrations from the contaminated aquifer, and (d) scanning electron microscopy (SEM) image showing etching and pitting of the mineral grains.

In another study through a vertical cross section of a hydrocarbon plume, Allen et al. documented that zones of higher bulk electrical conductivity were coincident with higher percentages of oil-degrading microbial populations. Using molecular microbial techniques determined by construction and evaluation of 16S rRNA clone libraries along a vertical gradient at the contaminated location, Allen et al. further showed that spatial heterogeneity in the microbial community structure and shifts in the microbial community were concomitant with vertical changes in bulk electrical conductivity. The shifts in the microbial communities were also related to the partitioning of the different phases of the hydrocarbon: methylotrophs and aromatic hydrocarbon degraders made up the dominant population in the residual phase; syntrophic populations characterized the dissolved phase, while hydrocarbon-degrading fermenters and iron and sulfur reducers made up the bulk of the populations within the free phase. More importantly, a multivariate statistical analysis suggested that elevated bulk electrical conductivity may be related to the presence and activities of specific microbial populations (e.g., syntrophic species α- and β-Proteobacteria). The results presented by Allen et al. suggest that geoelectrical techniques may help guide sampling in microbial ecology studies.

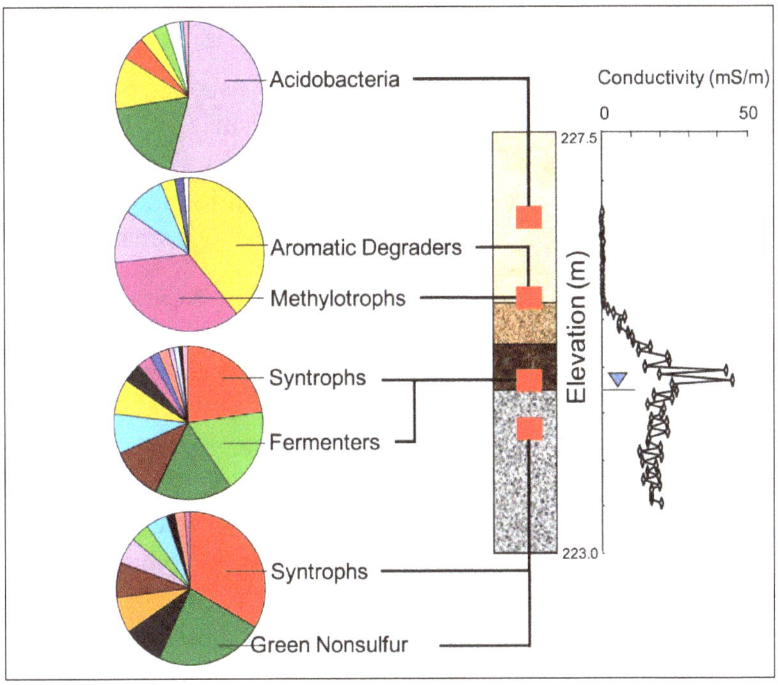

The vertical light nonaqueous phase liquid (LNAPL) distribution at a contaminated field site along with the corresponding bulk electrical conductivity measurements. The microbial community structures for samples taken from within the vadose, residual LNAPL, free phase LNAPL, and water-saturated zones as determined by 16S rRNA gene sequencing show significant differences.

Other metabolic by-products produced by micro-organisms include biogenic gases (e.g., CO_2, CH_4, H_2S, and H_2). These gases can impact geophysical properties. For example, methane production by archea in anaerobic peat soils can result in extensive free phase gas production in peatlands that has been noninvasively detected using GPR and variants of the technique in the laboratory.

Microbe-mediated Redox Processes

Under aerobic conditions bacteria first utilize oxygen as TEAs whereas in anaerobic environments other electron acceptors, such as sulfate in the presence of sulfate-reducing bacteria or Fe(III) in the presence of iron-reducing bacteria, are utilized. Microbial respiration consequently results in reduced conditions, and strong redox gradients in the Earth typically develop in the presence of bacteria. In this section we consider how microbe-induced redox gradients may drive detectable geophysical signatures.

Self potential is the only geophysical method dependent, as shown by theory, on redox gradients and the only method therefore to concern ourselves with in this section. In the self-potential (SP) method the electric fields generated by internal electric current sources in the Earth are recorded with a pair of precision electrodes carefully constructed, using a porous pot construction, so as to avoid reactions between the metallic electrode and the electrolyte. This is achieved by immersing metal electrodes in saturating solutions/gels of the metal (an identical solution being used for both electrodes making up the pair such that the redox state at the two metal surfaces is identical). Interest in the application of the SP method for indirectly detecting microbial activity from redox gradients produced has been fueled by a number of studies showing strong SP signals (exceeding 0.1 V) at sites where microbial degradation of hydrocarbons is occurring.

However, understanding of self-potential signals in the presence of microbial activity has been perhaps hindered by a number of factors including (1) interpretation using electrochemical models physically inconsistent with SP measurements and (2) nomenclature, whereby some authors have chosen to refer to measurements of redox chemistry occurring at an electrode (i.e., a reaction between the electrolyte and the metal electrode surface) as a "self potential". There are numerous well-understood mechanisms generating sources of electric current in the Earth, e.g., the electrokinetic source term resulting from the transport of excess charge in the electrical double layer at pore surfaces as a result of the viscous drag exerted by fluid flow through the pores. However, there are only two established current source mechanisms that can potentially generate self-potential signals in the presence of microbe-induced redox gradients: (1) electrodiffusional effects due to gradients in the chemical potentials of the charge carriers causing the redox gradient and (2) redox reactions and coupled electron transport associated with electronic conductors (geobattery effects).

Maineult et al. conducted a detailed laboratory study of electrodiffusional effects in the presence of sharp redox gradients produced by parallel advecting $KMnO_4$ and $FeCl_2$ solutions in contact along a reaction front. One of the incentives of this work was to ascertain whether an SP source mechanism additional to the electrodiffusional and electrokinetic terms was needed to satisfy SP signals observed across the redox front. They showed that despite the presence of a redox reaction occurring across this front (visible precipitation of $Fe(OH)_3$), the self potentials across this redox gradient were small (<0.01 V) and not significantly greater than those resulting from the combination of electrodiffusional and electrokinetic effects outside of the reaction front. They concluded that the SP signal corresponding to this second front is not related to the variation of the redox potential itself but rather to electrodiffusional mechanisms associated with changes in composition and concentration of the solution resulting from the redox reaction. This work suggests that (1) an alternative SP source mechanism is needed to explain the SP signals recorded at sites where biodegradation is occurring as the signals are large (>0.01 mV) and (2) this mechanism is absent in the abiotic experiments of Maineult et al.

The fact that an alternative mechanism is needed is evident from the work of Minsley et al. who inverted borehole SP signals (approximately normally distributed between 0 and 0.15 V) from a site of known dense nonaqueous phase liquid contamination for a distribution of current sources that could explain the measured potentials. They showed that the distribution of current sources obtained from the inversion was empirically related to the distribution of redox potential. Minsley et al. describe a conceptual model for the generation of the current sources associated with the biodegradation of contaminants (primarily perchloroethylene, trichloroethylene, and trichloroethane at their site) by microbe-mediated redox reactions. They suggest that electron transport, via the production of a chain of intermediate compounds until the final electron acceptor is reduced, must be balanced by an external current flow in the Earth such that the total electric current density is divergence-free. They present an analogy to the corrosion of two metals in a short-circuited energy-producing electrochemical cell whereby the electron transport via the short circuit is associated with the electron transfer involved in the reduction chain associated with the biodegradation. The driving force of this electrochemical cell was assumed to be the redox gradients caused by the biodegradation reactions occurring at the site.

Such an energy producing cell is the basis of the traditional "geobattery model" proposed to explain very large (~1.0 V) SP signals repeatedly recorded over electronically conductive ore bodies straddling the large redox gradient provided by the water table. The traditional geobattery for an ore body considers both the distribution of the redox potential and the kinetics of the chemical reactions at the surface of the metallic particles. The ore body participates directly in the two half-cell reactions of the electrochemical cell, consisting of anodic (oxidizing) and cathodic (reducing) reactions located at the bottom and the top of the ore body, respectively. The deep anodic reaction corresponds to the galvanic corrosion of the metallic body, whereas the cathodic reaction is the reduction of oxygen radicals at the top of the body. The driving force of the electrochemical cell is atmospheric oxygen dissolved in groundwater and the electronically conductive ore body serves as the short circuit for electron transport. This model has been widely used in mineral exploration to explain SP signals over metallic ore bodies.

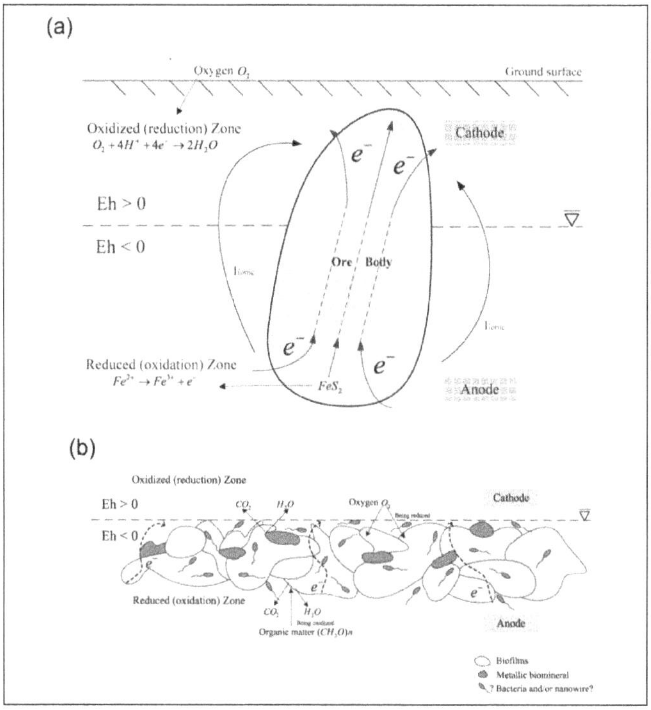

(a) The Sato and Mooney geobattery model whereby the redox gradient is attributed to the gradient of dissolved oxygen in groundwater and the ore body serves as the electron conductorand (b) the biogeobattery model proposed by Naudet et al. whereby the redox gradient is attributed to the gradient in dissolved Fe^{2+} resulting from oxidation of organic matter and biofilms at the water table serve as the electron conductor.

In order to explain the large (~0.4 V) SP signals recorded over the Entressen Landfill that could not be explained by electrodiffusional effects, Naudet et al. devised an alternative biogeobattery model as shown in figures. Here the distribution of redox potential is associated with a strong redox gradient between highly reducing conditions below the water table within the plume (due to biodegradation and oxygen depletion) and the oxidized zone above the water table. The driving force of this electrochemical cell is dissolved Fe^{2+} resulting from the oxidation of organic matter by bacteria followed by reduction of Fe(III) oxides. To complete the geobattery, Naudet et al. proposed that biofilms forming at the groundwater table could possibly provide the short circuit and required electron transport. This fascinating model has generated considerable interest and debate in the field of biogeophysics as it opens intriguing possibilities for monitoring microbial processes from large SP signals associated with geobattery effects.

A weakness of this model is that the ability of biofilms to facilitate electron transport, for example, over the scale of the groundwater interface, is currently unproven. However, there is growing evidence that metal-reducing organisms, such as Shewanella and Geobacter, produce electrically conductive appendages called bacterial nanowires that facilitate electron transfer to solid phase electron acceptors. In Geobacter, nanowires are composed of a unique pilin protein. These "geopili" are electrically conductive, presumably in the absence of traditional electron transport proteins such as cytochromes. In Shewanella, nanowires require multiheme cytochromes to be electrically conductive, as evidenced by the fact that mutants lacking genes for a pair of extracellular decaheme cytochromes (MtrC and OmcA) produced nonconductive appendages. However, the complete composition and mechanisms of electron transfer in nanowires produced by either Shewanella or Geobacter remain ill defined. We therefore consider the assumption that electron transport is occurring at sites of biodegradation via nanowires forming at the groundwater surface uncertain.

The pili of Geobacter have been documented to play an important role in not only enhancing biofilm growth but also in forming an electronically conductive network within biofilm, enhancing electricity generation. The use of nanowires as a potential mechanism for electron transfer at organic-rich sites is not entirely impossible, and the contention is not whether the bacteria can produce a network of nanowires linking cells in a biofilm; rather, the debate and surrounding controversy are over whether such a network of nanowires within biofilm can allow for direct cell-to-cell long-range electron transfer at the spatial (e.g., centimeter) scales suggested by the Naudet et al. model. There remains a large knowledge gap to be bridged between the scale at which electron transport by geopili in metal-reducing bacteria has been proven (nanometer scale) and the scale of the representative elementary volume sampled with a geophysical measurement (centimeter to meter scale).

Clearly, controlled laboratory experiments are needed to prove that biogeobattery effects exist in microbial systems and can generate large SP signals as recorded at field sites. Ntarlagiannis et al. performed column experiments, where scanning electron microscopy (SEM) revealed geopili linking cells to cells and cells to mineral surfaces, leading them to suggest that the column was effectively electrically "hardwired". Given the large redox gradient across the column, the essential elements of a geobattery were produced. Interestingly, they recorded large (~0.6 V) electrical potentials on Ag-AgCl metal electrodes straddling the column, which were interpreted as SP signals arising from a biogeobattery. Perhaps most importantly, these geophysical data were presented as evidence that geopili could facilitate electron transport (hardwiring) at the column scale. However, Ntarlagiannis et al. used Ag-AgCl metal electrodes that were in contact with the column electrolyte and thus sensitive to the sum of contributions from (1) self potentials associated with any current source in the column (i.e., the biogeobattery) and (2) open circuit galvanic cell (GC) potentials between the electrodes resulting from the tendency of the electrode surface to react with the electrolyte as a result of any difference in redox state at the two electrodes. As such, GC potentials due to local reactions at the electrode (henceforth termed electrodic potentials) are very sensitive to changes in redox state between the two electrodes and can exceed 0.6 V in the presence of metal-reducing organisms. The results reported by Ntarlagiannis et al. therefore warrant further studies and validation.

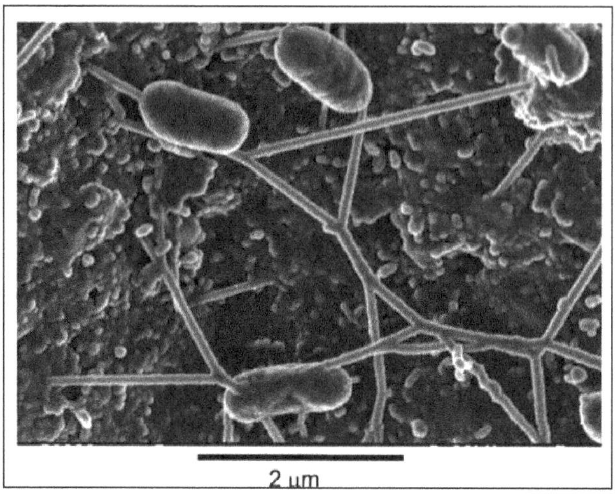

SEM images revealing geopili linking cells to cells and cells to mineral surfaces, suggesting electrical “hardwiring” of a soil column as postulated by Ntarlagiannis et al.

We note that some geophysicists have intentionally explored the use of electrodic potentials for monitoring activity of sulfate reducing microbial communities under anaerobic conditions. Metallic electrodes in contact with the pore-filling electrolyte are routinely used for electrochemical applications to measure geochemical parameters, such as Eh and pH, as well as specific chemical concentrations (e.g., Br using a bromide-specific electrode). A common target compound is sulfide, which can be accurately monitored using Ag-AgS electrodes. This GC measurement exploits the known reactivity between the target compound and the electrode used, with the resulting electrodic potential usually being measured relative to an electrode with a fixed potential.

Williams et al. showed how a pair of Ag-AgCl metal electrodes, deployed in a fashion similar to a conventional geophysical survey, records a GC potential when they straddle a gradient in bisulfide (HS^-) concentration. Oxidation of the Ag-AgCl anode results from a half-cell reaction in the presence of elevated HS^- concentrations relative to the cathode, whereas reduction of the AgCl electrode coating occurs at the cathode. The anode effectively becomes an Ag/Ag_2S electrode with the known sensitivity to HS^-. A voltmeter connected between these electrodes will record the open circuit potential caused by the differences in HS^- concentration at the local fluid-electrode interface between the two electrode locations. This open circuit potential is a measure of the tendency of a spontaneous reaction between these electrodes to proceed (significant current flow does not occur because of the high input impedance of the voltmeter), and it increases with an increase in the HS concentration difference between the two electrodes. These electrodic potentials have been used to observe microbial sulfate reduction in saturated synthetic soils following amendment with organic carbon and in natural wetland soils and to provide auxiliary data in complex resistivity studies of FeS biomineralization. Although such measurements are diagnostic of microbial sulfate reduction, they only detect redox chemistry at the electrode surface and are thus an electrochemical technique and not a biogeophysical technique; for example, they can provide no information on microbial activity occurring in the Earth from sensors placed at the surface. It is perhaps inappropriate then to describe such GC measurements as SP measurements. As stated in section and explicitly clarified by Slater et al., the SP geophysical technique is fundamentally concerned with the measurement of potentials resulting from current sources in the Earth that can only be probed using electrodes carefully designed to eliminate electrochemical effects at the surface of the metallic part of the electrode.

It is clear that there is considerable uncertainty, and perhaps confusion, regarding the source of SP signals recorded at sites of active microbial degradation. Certainly, some of the confusion has been driven by a misunderstanding of the basis of the SP measurement and the associated requirements of an SP electrode. The proposed biogeobattery model presents an elegant explanation for microbe-produced SP signals, but suggestions that geopili in biofilms could facilitate the required electron transport require the support of more robust data sets. Arora et al. did also suggest that precipitation of metallic particles within the vicinity of the water table could provide the electron transport required to complete the biogeobattery model of Naudet et al.

Biogeochemical (Microbe-mineral) Transformations

In this section we turn our attention to how the microbial modification of mineral surfaces impacts geophysical measurements. Microbial processes can result in both precipitation of new minerals onto existing mineral surfaces and dissolution of existing mineral surfaces, e.g., by organic acids produced as an end result of biodegradation of organic compounds. Here we assess the significance of geophysical measurements conducted during both microbe-induced precipitation and dissolution of mineral surfaces.

Biogeochemical processes resulting in mineral precipitation fundamentally alter the physicochemical properties of the interface and, when microbe-induced precipitation is extensive, may also affect the volumetric physical properties of porous media. Table summarizes three of the most well known microbial enzymatic reactions and the associated precipitation of secondary minerals. We consider here the evidence suggesting that these reactions can be noninvasively detected with geophysical methods.

Table: Microbial enzymatic reactions and the precipitation of secondary minerals.

Bacteria	Reactions
Ureolytic bacteria (e.g., Bacillus pasteurii)	$NH_7CONH_2 + 2\ H_2O \leftrightarrow 2\ NH_4^+ + CO_3^{2-}$ $Ca^{2+} + CO_3^{2-} \leftrightarrow CaCO_{3(s)}$
Sulfate reducing bacteria (e.g., Desulfovibrio desulfuricans)	$2\ CH_3CHOHCOO^- + SO_4^{2-} + 2\ H^+ \leftrightarrow 2\ CH_3COO^- + 2\ CO_2 + 2\ H_2O + HS^-$ $Fe^{2+} + HS^- \leftrightarrow FeS_{(s)} + H^+$

Iron(III) reducing bacteria (e.g., Shewanella putrefaciens)	$CH_3CHOHCOO^- + 4Fe(OH)_3 + 7H^+ \leftrightarrow 4 Fe^{2+} + CH_3COO^- + HCO_3^- + 10H_2O$ $Fe^{2+} + HCO_3^- \leftrightarrow FeCO_{3(s)} + H^+$ $2\ Fe(OH)_3 + Fe^{2+} \leftrightarrow Fe_3O_4 + 2H_2O + 2H^+$ $Fe(OH)_{3(s)} \xrightarrow{Fe(II)} FeOOH_{(s)} + H_2O$

Microbe-mediated Precipitation of Metallic Iron Minerals

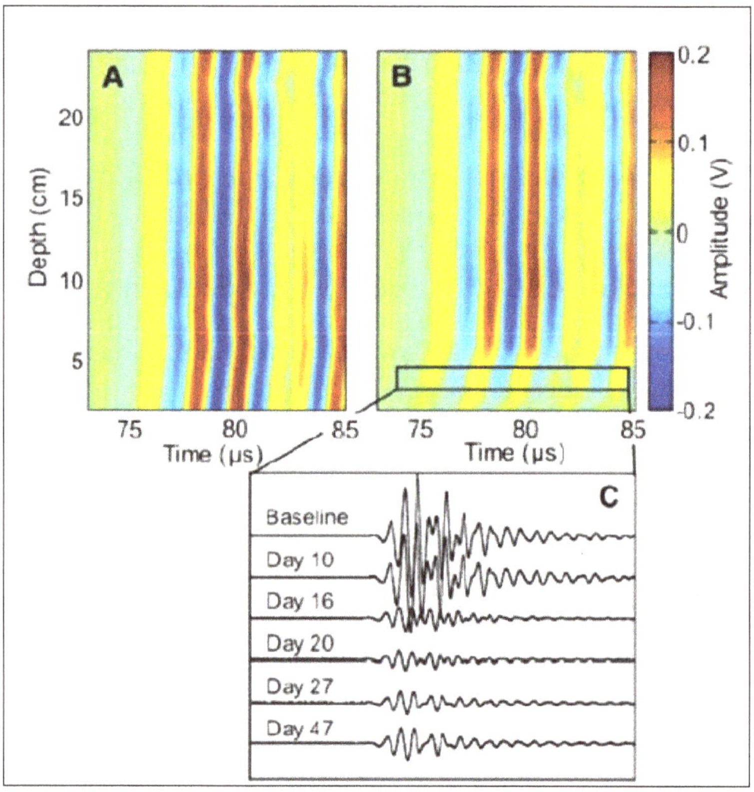

Experimental evidence of attenuation of seismic waves as a result of biomineral formation: (a) baseline waveform prior to biomineralization, (b) waveform after 20 days showing attenuation toward bottom of column in zone of observed biomineralization, and (c) time variations in wave pulse at 2 cm from bottom of column.

We first consider the microbial precipitation of iron minerals. The biological precipitation of these minerals plays a fundamental role in regulating the cycling of iron, sulfur, and carbon on the Earth's surface. In anaerobic environments, iron-reducing bacteria and sulfate-reducing bacteria can use Fe(III) and sulfate, respectively, as terminal electron acceptors. Ferrous iron produced by iron-reducing bacteria can promote the precipitation of secondary minerals such as siderite ($FeCO_3$), magnetite (Fe_3O_4), and goethite (FeOOH). Sulfide produced during microbial sulfate reduction can react with iron(II) produced by iron-reducing bacteria to precipitate iron sulfide minerals. Sedimentary sulfide minerals can act as diagenetic cements and represent an important sink for biogenic sulfide. Sulfate-reducing bacteria account for ~50% of total carbon mineralization in active marine sediments. In hydrocarbon-contaminated groundwater aquifers, the microbial reduction of Fe(III) and sulfate has been linked to the biological precipitation of siderite ($FeCO_3$), pyrite (FeS_2), mackinawite (FeS), and amorphous iron sulfide (FeS). These microbe-induced precipitation mechanisms regulate the cycling of iron, sulfur, and carbon and may be essential to organic contaminant biodegradation and the remediation of mine wastes using natural/artificial wetlands.

The need for new technologies capable of remotely monitoring bioremediation processes associated with sulfate-reducing communities at contaminated sites stimulated a series of experiments to determine the ability of geophysical measurements to track the evolution of metal sulfide biominerals in laboratory columns. Williams et al. showed that electrical and seismic signatures were generated as a result of microbe-induced ZnS and FeS precipitation within saturated sand-packed columns using the sulfate-reducing bacterium Desulfovibrio vulgaris. Observed decreases in seismic wave amplitude throughout the experiment were attributed to the development of differential elastic moduli resulting from the accumulation of metal sulfide-encrusted microbes within the pore space.

The low-frequency complex electrical conductivity data showed a more complicated response, the column becoming polarizable in the early stages of the experiment but transitioning to conductive in the later stages of the experiment. Williams et al. attributed the early response to the initial formation of dispersed metal sulfide encrusted biominerals presenting a net high metal-fluid interfacial area. The subsequent loss of polarizability and conduction was attributed to either (1) the formation of biofilm polymers that could conceivably screen metal sulfides from the pore fluids (rendering them unpolarizable) or (2) formation of electrically conductive short circuits along grain surfaces accompanying the aging of metal precipitates. Ntarlagiannis et al. further analyzed the complex conductivity (at a single frequency) results from a series of measurement points along the column and concluded that the temporal behavior of the electrical signals could be explained in terms of (1) the amount of metal-fluid interfacial area and (2) the amount of electronic conduction resulting from microbial growth, chemotactic movement, and final coagulation of biominerals. Interestingly, they found that the polarization peak correlated closely in time with the peak in the microbial activity rates inferred from lactate concentration data, suggesting a common metal surface area effect. Furthermore, the onset of microbial activity was deduced from the onset of polarization observed with the electrical measurements.

Further experiments on FeS biomineralization produced by Desulfovibrio vulgaris under anaerobic conditions, followed by subsequent biomineral dissolution upon return to an aerobic state, have attempted to quantify changes in pore geometry resulting from biomineral formation by modeling the frequency response of complex conductivity data. Slater et al. describe a column experiment similar to Williams et al. , whereby parameters extracted from Cole-Cole relaxation modeling of the complex conductivity data exhibit behavior conceptually consistent with increases in FeS surface area and pore size reduction during biomineral growth and subsequent decreases in FeS surface area and pore expansion accompanying biomineral dissolution. This effort was further extended by Personna et al. who applied an empirical relation, linking the polarization magnitude to specific surface area per unit pore volume (S_{por}) of the sand-metal mixtures, to estimate changes in S_{por} driven by FeS precipitation and subsequent dissolution.

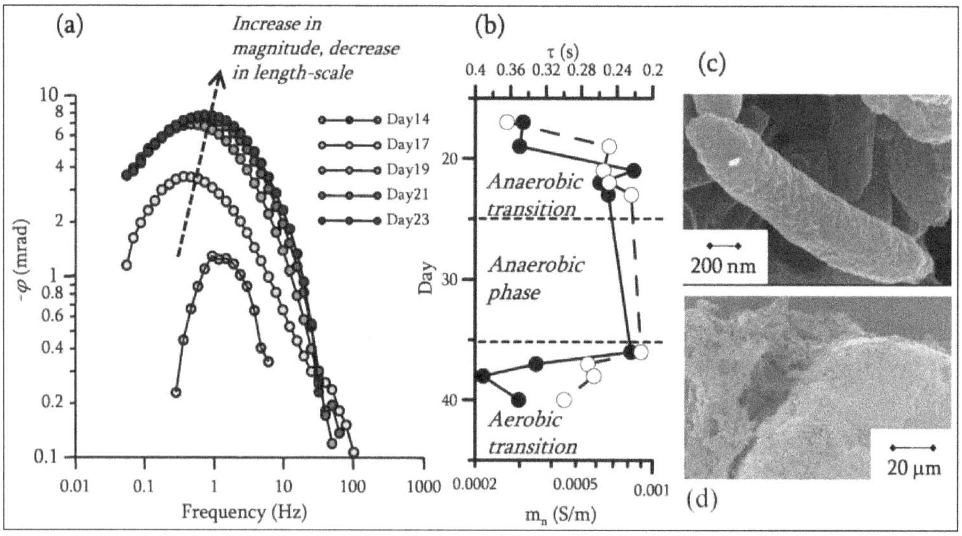

(a) Polarization spectra as a function of time during FeS biomineralization induced by an anaerobic transition. (b) This biomineralization response is largely reversible during subsequent dissolution (aerobics transition) as apparent from the behavior of modeled Cole-Cole parameters (normalized chargeability, m_n, and time constant, τ). SEM images for samples extracted from the column on termination of experiment showing (c) tube-like, elongate D. vulgaris biominerals and (d) quartz sand particle encrusted with biomineralization.

The proven sensitivity of complex conductivity measurements to the formation, movement, and dissolution of FeS biominerals is perhaps unsurprising given the profound impact that electronically conductive metallic minerals have on the electrical properties of rocks. In fact, the induced polarization and spectral induced polarization geophysical prospecting methods were largely developed for locating buried disseminated FeS deposits. This biogeophysics research has confirmed the existence of measurable complex conductivity signals associated with the formation of sulfide minerals by sulfate-reducing bacteria under anaerobic conditions, suggesting that these established geophysical techniques could potentially be adapted to monitoring of microbe-mineral transformations in the Earth.

Microbe-mediated Precipitation of Nonmetallic Minerals

Microbial processes may also result in the precipitation of semimetallic or nonmetallic minerals. One important microbial enzymatic reaction is the microbial precipitation of calcite, due to its ability to sequester metal contaminants and its impact on the hydraulic properties of the subsurface. Numerous micro-organisms have been shown to induce the precipitation of calcite. Microbial calcite precipitation is mediated by metabolically active bacteria increasing the saturation state and serving as heterogeneous nucleation substrates for mineral formation. One example of metabolically induced calcite precipitation is bacterial hydrolysis of urea. The production of ammonia and dissolved inorganic carbon during enzymatic decomposition of urea increases pH and alkalinity, both of which promote calcite precipitation. Calcite precipitation induced by ureolytic bacteria has been shown to sequester mobile radionuclide elements such as Sr^{90}. Furthermore, calcite precipitation induced by bacteria has been shown to form cements in porous media and affect subsurface fluid flow. Microbial oxidation of organic matter produces dissolved inorganic carbon and leads to the accumulation of carbonate and bicarbonate ions along the fluid flow path. Once the groundwater becomes supersaturated with respect to $CaCO_3$(s), calcite will precipitate to form intergranular calcite cements.

We hence consider here the possible geophysical detection of microbe-mediated precipitation of semimetallic or nonmetallic mineral phases. Whereas the electrical properties of conductive mineral phases support detection using electrical geophysics methods, the electrical geophysical detection of nonmetallic mineral phases associated with microbial processes is intuitively less certain. Instead of exploiting the conductivity/polarization enhancement associated with the formation of metallic iron minerals we must rely on the intuitively small electrical signals associated with changes in pore volume/pore tortuosity and/or surface area/surface roughness resulting from the precipitation of nonmetallic minerals. In contrast, there are profound changes in the elastic properties of soils and rocks (particularly those affecting shear velocities) during precipitate formation.

Microbe-induced precipitation of calcite is currently being explored as a novel strategy for improving the engineering properties (e.g., soil strength, permeability, and compressibility) of soils. Here ureolysis is stimulated to form calcite cement that acts to stiffen the soil matrix. DeJong et al. examined calcite precipitation by the common soil micro-organism, Bacillus pasteurii, using SEM to verify the formation of a cemented sand matrix with a concentration of precipitated calcite forming bonds (verified with X-ray compositional mapping) at particle-particle contacts. Shear waves are very well suited to monitoring changes in the particle soil matrix due to precipitation as shear velocity (V_s) is largely unaffected by pore fluid composition and directly dependent on void ratio, coordination number (average number of surrounding particles a given particle is in contact with), and confining stress. DeJong et al. provide a conceptual model of how changes in V_s will detect biomediated calcite precipitation. The process shows a sequence of discrete injections of urealytic micro-organisms. During injection V_s remains constant as the pore fluid is replaced. Biological activity then gradually raises the pH to a level where calcite precipitation occurs. The initial microbial treatment resulted in a strong initial increase in V_s believed to be caused by initial binding of the microbes to the soil matrix. The velocity increases until some limiting condition is reached; for example, no additional calcite is available in solution. V_s then remains constant until a subsequent injection occurs and the process is repeated. Experimental results of this process are shown in figure, with each vertical arrow indicating an additional discrete injection. Shear wave velocity shows a total 270% increase due to ureolysis and formation of calcite cement, suggesting that the measurement of shear wave velocities is an excellent technology for monitoring microbial ureolysis.

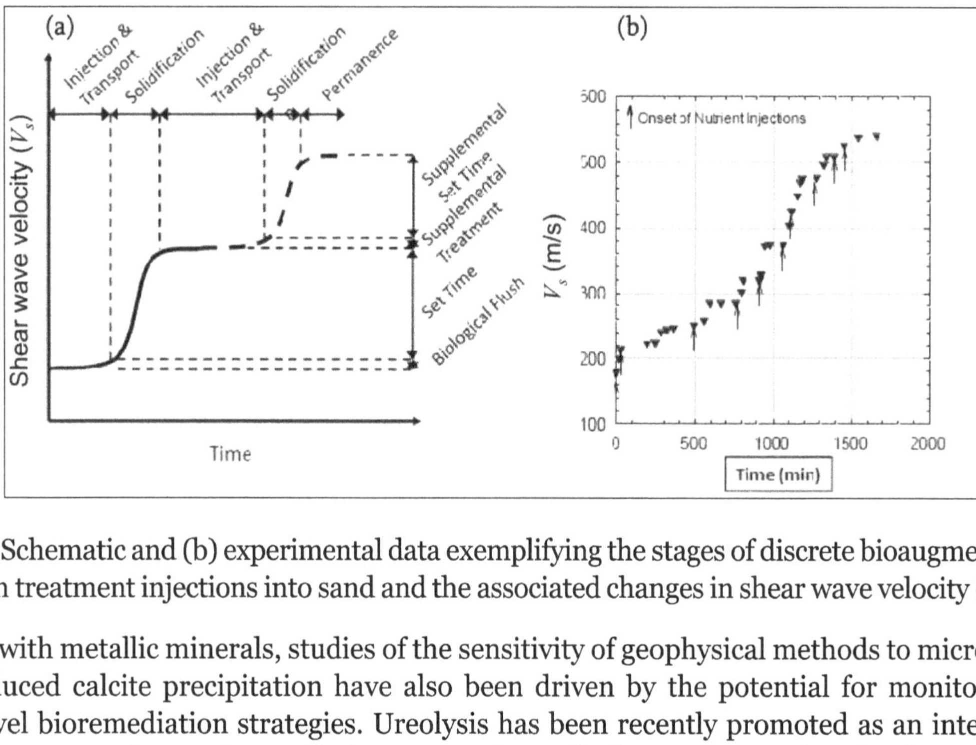

(a) Schematic and (b) experimental data exemplifying the stages of discrete bioaugmentation treatment injections into sand and the associated changes in shear wave velocity (V_s).

As with metallic minerals, studies of the sensitivity of geophysical methods to microbe-induced calcite precipitation have also been driven by the potential for monitoring novel bioremediation strategies. Ureolysis has been recently promoted as an integral component of a novel remediation approach for divalent trace metal and radionuclide contaminants whereby the contaminants are immobilized by coprecipitation in calcite. However, recent work to monitor calcite precipitation by ureolytic bacteria using complex resistivity measurements found that the conduction and polarization signals are too small to be reliably indicative of the precipitation process.

Magnetic Minerals/Metal Cycling

In addition to mediating the precipitation of different mineral phases microbial activity can also result in the alteration of mineral phases potentially impacting the magnetic properties of the sediments. A study by Carlut et al.examined the impact of bacterial activity on magnetic properties of young ocean floor basalts. Fresh basalt samples obtained from young pillow lavas at different locations along the East Pacific Ridge were exposed to the action of the sulfate reducing bacteria Desulforvibrio desulfuricans for a total of 355 days during which time natural remnant magnetization measurements were monitored. The results showed that most magnetized samples lost ~30% of their original magnetization. SEM imaging and electron dispersive analysis revealed the cycling of iron mineral phases from titanomagnetites which is responsible for the original magnetization of the ocean floor basalts to less magnetic iron sulfide phases. Carlut et al. concluded that micro-organisms could play an important role in the rapid loss of magnetization of near-surface oceanic crust at ridge axis and could be responsible for the globally observed central magnetic high along ridge axis. The above results suggest that established magnetic geophysical techniques could be used to investigate the activity of bacteria on the deep ocean floor particularly as it relates to the biotransformation of magnetic mineral phases.

Near-surface Geophysics

Automatic ground penetrating Radar (upGPR).

Near-surface geophysics is the use of geophysical methods to investigate small-scale features in the shallow (tens of meters) subsurface. It is closely related to applied geophysics or exploration geophysics. Methods used include seismic refraction and reflection, gravity, magnetic, electric, and electromagnetic methods. Many of these methods were developed for oil and mineral exploration but are now used for a great variety of applications, including archaeology, environmental science, forensic science, military intelligence, geotechnical investigation, treasure hunting, and hydrogeology. In studies of the solid Earth, the mean feature that distinguishes geophysics from geology is that it involves remote sensing. Various physical phenomena are used to probe below the surface where scientists cannot directly access the rock. Applied geophysics projects typically have the following elements: data acquisition, data reduction, data processing, modeling, and geological interpretation. This all requires various types of

"Geophysical surveys". The various Geophysical surveys one can carry out are, Gravity, Magnetic, Seismic, Magneto-telluric etc.

Data Acquisition

A geophysical survey is a set of measurements made with a geophysical instrument. Often a set of measurements are along a line, or traverse. Many surveys have a set of parallel traverses and another set perpendicular to it to get good spatial coverage. Technologies used for geophysical surveys include:

- Seismic methods, such as reflection seismology, seismic refraction, and seismic tomography.
- Seismoelectrical method.
- Geodesy and gravity techniques, including gravimetry and gravity gradiometry.
- Magnetic techniques, including aeromagnetic surveys and magnetometers.
- Electrical techniques, including electrical resistivity tomography, induced polarization and spontaneous potential.
- Electromagnetic methods, such as magnetotellurics, ground penetrating radar and transient/time-domain electromagnetics.
- Borehole geophysics, also called well logging.
- Remote sensing techniques, including hyperspectral.

Data Reduction

The raw data from a geophysical survey must often be converted to a more useful form. This may involve correcting the data for unwanted variations; for example, a gravity survey would be corrected for surface topography. Seismic travel times would be converted to depths. Often a target of the survey will be revealed as an anomaly, a region that has data values above or below the surrounding region.

Data Processing

The reduced data may not provide a good enough image because of background noise. The signal-to-noise ratio may be improved by repeated measurements of the same quantity followed by some sort of averaging such as stacking or signal processing.

Modeling

Once a good profile is obtained of the physical property that is directly measured, it must be converted to a model of the property that is being investigated. For example, gravity measurements are used to obtain a model of the density profile under the surface. This is called an inverse problem. Given a model of the density, the gravity measurements at the surface can be predicted; but in an inverse problem the gravity measurements are known and the density must be inferred. This problem has uncertainties due to the noise and limited coverage of the surface, but even with perfect coverage many possible models of the interior could fit the data. Thus, additional assumptions must be made to constrain the model. Depending on the data coverage, the model may only be a 2D model of a profile. Or a set of parallel transects may be interpreted using a 2½D model, which assumes that relevant features are elongated. For more complex features, a 3D model may be obtained using tomography.

Geological Interpretation

The final step in a project is the geological interpretation. A positive gravity anomaly may be an igneous intrusion, a negative anomaly a salt dome or void. A region of higher electrical conductivity may have water or galena. For a good interpretation the geophysics model must be combined with geological knowledge of the area.

Magnetic Surveying

Magnetic surveying can be done on a planetary scale (for example, the survey of Mars by the Mars Global Surveyor) or on a scale of meters. In the near-surface, it is used to map geological boundaries and faults, find certain ores, buried igneous dykes, locating buried pipes and old mine workings, and detecting some kinds of land mines. It is also used to look for human artifacts. Magnetometers are used to search for anomalies produced by targets with a lot of magnetically hard material such as ferrites.

Microgravity Surveying

High precision gravity measurements can be used to detect near surface density anomalies, such as those associated with sinkholes and old mine workings, with repeat monitoring allowing near-surface changes over these to be quantified.

Ground Penetrating Radar

Ground penetrating radar is one of the most popularly used near-surface geophysics in forensic archaeology, forensic geophysics, geotechnical investigation, treasure hunting, and hydrogeology, with typical penetration depths down to 10m below ground level, depending upon local soil and rock conditions, although this depends upon the central frequency transmitter/receiver antennae utilised.

Bulk Ground Conductivity

Bulk ground conductivity typically uses transmitter/receiver pairs to obtain primary/secondary EM signals from the surrounding environment (note potential difficulty in urban areas with above-ground EM sources of interference), with collection areas depending upon the antennae spacing and equipment used. There are airborne, land- and water-based systems currently available. They are particularly useful for initial ground reconnaissance work in geotechnical, archaeology and forensic geophysics investigations.

Electrical Resistivity

Electrical resistivity tomography profile.

The reciprocal of conductivity, electrical resistivity surveys measure the resistance of material (usually soil) between electrical probes, with typical penetration depths 1-2x the electrode separations. There are various electrode configurations of equipment, the most typical using 2 current and 2 potential electrodes in a dipole-dipole array. They are used for geotechnical, archaeology and forensic geophysics investigations and have better resolution than most conductivity surveys. Note they do experience significant changes with soil moisture content, a difficulty in most site investigations with heterogeneous ground and differing vegetation distributions.

Tectonophysics

Tectonophysics, a branch of geophysics, is the study of the physical processes that underlie tectonic deformation. The field encompasses the spatial patterns of stress, strain, and differing rheologies in the lithosphere and asthenosphere of the Earth; and the relationships between these patterns and the observed patterns of deformation due to plate tectonics.

Tectonophysics is concerned with movements in the Earth's crust and deformations over scales from meters to thousands of kilometers. Examples of such processes include mountain building, the formation of sedimentary basins, postglacial rebound of regions such as Fennoscandia, plate tectonics, volcanoes and earthquakes. This involves the measurement of a hierarchy of strains in rocks and plates as well as deformation rates; the study of laboratory analogues of natural systems; and the construction of models for the history of deformation.

Tectonophysics was adopted as the name of a new section of AGU on April 19, 1940 at AGU's 21st Annual Meeting. According to the AGU website using the words from Norman Bowen, the main goal of the tectonophysics section was to "designate this new borderline field between geophysics, physics and geology for the solution of problems of tectonics." Consequently the claim below that the term was defined in 1954 by Gzolvskii is clearly incorrect. Since 1940 members of AGU had been presenting papers at AGU meetings, the contents of which defined the meaning of the field.

Tectonophysics was defined as a field in 1954 when Mikhail Vladimirovich Gzovskii published three papers in the journal Izvestiya Akad. Nauk SSSR, Sireya Geofizicheskaya: "On the tasks and content of tectonophysics", "Tectonic stress fields", and "Modeling of tectonic stress fields". He defined the main goals of tectonophysical research to be study of the mechanisms of folding and faulting as well as large structural units of the Earth's crust. He later created the Laboratory of Tectonophysics at the Institute of Physics of the Earth, Academy of Sciences of the USSR, Moscow.

Hydrogeophysics

The field "hydrogeophysics" emerged in the 1990s as a multi-disciplinary subject that focuses on the use of geophysical methods for characterising subsurface features, determining hydrogeological properties and monitoring processes relevant to soil and groundwater processes. Hydrogeophysical methods can allow large scale aquifer characterization, previously unobtainable through conventional hydrogeological techniques. In addition, time-lapse deployment of appropriate methods can give useful insight into complex subsurface processes, aiding hydrological model development and the assessment of groundwater restoration strategies.

Hydrology has traditionally relied upon the availability of point measurements: precipitation at rain gauges, groundwater potentiometric surface at boreholes and, more recently, soil moisture content as measured, for example, by time-domain reflectometry. This body of information is necessary, but often far from being complete. In particular, it has been largely demonstrated that, given the point information above, there will always be a number of different models that reproduce equally well the observed hydrological data (such as river discharge). This equifinal nature of different models is limited to their capability of matching the observed limited and local data, and may not be reflected in their prediction capability, i.e., the model forecasts may differ substantially should the forcing conditions (e.g., precipitation) be changed beyond the observed values. This critical model limitation is essentially driven by the fact that the actual structure of the subsurface is poorly known in terms of geometry, geology and hydraulic properties; this limitation often forces the models to be black boxes, either explicitly or implicitly due to the poor knowledge of the actual system characteristics.the models, but rather requires extra information. This is one of the fundamental needs addressed by hydrogeophysical measurements.

Hydrology research has developed a number of sophisticated tools to try and account for model uncertainty, particularly with respect to the limited knowledge of the subsurface in terms of its hydraulic property distribution. In particular, stochastic techniques have had a tremendous development in both surface and subsurface hydrology. However, stochastic models require even more data than traditional deterministic models, as an estimate of the underlying spatial statistical structure of the governing parameters is needed. As a consequence, during the 1990s there was a rapid growth in the use of geophysics to try and provide spatially dense, quantitative information about hydrological properties and processes. This need, in addition to the growing availability of fast field acquisition instruments and powerful computational tools, has led to much of the current developments in hydrogeophysics.

Important new requirements have been posed lately upon hydrological characterization and modelling, for example, by the new European Union regulations (in particular the Water Framework Directive 2000/60/EC, and the Groundwater Directive 2006/118/EC). In particular, the growing issue of water quality requires that the presence and location of contaminants be assessed, and their migration in the hydrological system be monitored. The limitations of hydrological modelling and point measurements are even more severe in the case of water quality than in water quantity assessment. Here too geophysical methods have a potential role.

Hydrogeophysics is asked to provide data for the following purposes:

- Structural characterization.
- Fluid-dynamics.
- Presence and motion of contaminants.

Geophysics has long been used to support hydrogeological studies, but mainly for lithological boundary delineation, i.e., to support structural (geometrical) characterization of aquifers. This use of geophysics does not fully exploit the geophysical measurements, but rather seeks to highlight contrasts in whatever physical property can distinguish one geological formation from its neighbours. This classical approach is also common to other major applications of geophysics, such as in the petroleum industry. However, geophysics has long been used also to exploit the physical nature of measurements, and translating these measurements into quantitative estimates of the soil/rock properties of interest. This has been particularly used for downhole measurements, i.e., for borehole logs. This translation requires that suitable constitutive laws link physical and structural properties of soil/rock, and the relevant discipline is petrophysics. The use of borehole logs for hydrogeological studies has a long history too even though it has never reached the popularity enjoyed in petroleum applications.

In view of the needs above, the current developments in hydrogeophysics are aimed at providing quantitative information on the hydrological and hydraulic characteristics of the soil and subsoil, as well as quantitative data on the presence and motion of fluids and solutes in and

Table: Use and effectiveness of geophysical methods for hydrological studies.

	Structure	Fluid dynamics	Contamination
Gravimetry	+	++	
Magnetics	+		
Seismics	+ +	+	
DC resistivity	+ +	+ +	+
Electromagnetics	+	+ +	+
Induced Polarization		+ +	+ +
Self Potential	+	+	+
GPR	+	+ +	+
NMR	+	+ +	

Out of the subsurface. Hydrogeophysics is therefore becoming a key instrument towards an effective characterization of hydrological systems.

Among the available geophysical methods, not all of them are equally suitable for hydrogeophysical applications. Each geophysical technique measures at least one physical quantity:

- Seismics: Elastic moduli and density.
- Gravimetry: Density.
- Magnetics: Susceptability and permanent magnetization.
- Geoelectrics (dc resistivity, e.g., electrical resistivity tomography (ert)): Electrical conductivity.
- Geoelectrics (induced polarization (ip)): Complex Electrical Conductivity.
- Electromagnetic (em) methods: Electrical Conductivity.
- Self potential (sp): Electrical conductivity and potential sources.
- Ground penetrating radar (gpr): Dielectric constant, electrical conductivity.
- Nuclear magnetic resonance (nmr): Number or protons, free pathway in pores, electrical conductivity.

In view of the hydrological aspects described above, potentially all geophysical methods can have some useful hydrogeophysical application, particularly in terms of structure characterization. However, some techniques have a more specific link to hydrological properties and to the presence/motion of water: these are listed above. In particular, the physical quantities more specifically affected by water presence or motion are, among the ones most commonly measured, electrical conductivity via ERT and dielectric constant via GPR. Other more specialized measurements (IP, SP and NMR) have strong connection to the pore-medium structure and presence/motion of water, solutes and free-phase contaminants, but are still of less common use as the relevant signal can be either below ambient noise or of uncertain attribution to hydrological causes.

Hydrological Problems and Hydrogeophysical Approaches

Effective hydrological studies require a quantitative assessment of water presence and motion in the environmental compartments of interest. Depending on the specific aims of the study at hand, one or more of the following compartments shall be considered: saturated zone, vadose (unsaturated) zone, shallow soil. The three compartments differ not only in terms of their hydrological role and mechanisms, but also in terms of the hydrogeophysical measurements that can be conducted profitably on them.

Saturated Zone

Aquifer studies are a traditional area of interest of subsurface hydrology. Their focus has been progressively shifting from water quantity to water quality problems. Transport of dissolved substances in groundwater is the most important mechanism controlling the migration of pollutants in the subsurface and is strongly controlled by geological heterogeneity at a variety of scales. In particular, hydraulic conductivity can vary by up to thirteen orders of magnitude. This fact has as a consequence a strong spatio-temporal variability of solute concentrations, that makes conventional monitoring techniques, based on a few boreholes and limited water sampling in space and time, often incapable of capturing the variability of transport properties, as well as the complexity of transport processes. Hydrogeophysics can provide spatially and temporally dense information on the evolution of solute plumes, particularly during tracer tests. Electrical and electromagnetic techniques have been predominantly used for these purposes, as they are sensitive to changes in aqueous phase electrical conductivity caused by saline tracers. Tomographic techniques offer the possibility to construct "images" of the subsurface, in 2D or 3D, that are well suited to picture the evolution of solute plumes. ERT, in particular, has been the key methodology applied for saline tracer imaging, but some notable examples using GPR attenuation for the same purpose have also been made possible, due to the increased electrical conductivity of the tracer. In all cases, and in contrast to structural hydrogeological characterization, where "static" properties of the subsurface are explored, transport characterization involves the monitoring of "dynamic" processes associated with spatio-temporal variations of subsurface state variables. The mapping and monitoring of transport processes therefore requires application of time-lapse geophysical methodologies that allow the user to distinguish between static and dynamic effects. Early applications were limited to imaging solute transport as accurately as possible, but did not provide estimates of hydraulic parameters and their spatial variability. In order to quantify the hydraulic parameters of the subsurface, it is essential to make use of hydrological models. The geophysical time-lapse data are used as equivalent concentration data to infer the timing and location of tracer breakthrough. In conjunction with transport models, such data can be directly interpreted in terms of transport parameters, such as flow velocity and dispersivity. The recent literature on time-lapse ERT applied to saline tracer tests follows this conceptual pathway to different extents. Tracer mass balance issues linked to ERT resolution issues prompted a series of possible approaches to tackle the limited resolution of ERT and the research is ongoing.

Vadose Zone

The vadose zone, i.e., the part of subsurface above the water table, is home to a number of key processes that control the mass and energy exchanges between the subsurface and soil surface. Vadose zone hydrology provides information about exchanges with the soil compartment, and from there with the atmosphere, and subsurface water migration, with strong implications in water resources management: aquifer recharge is controlled by movement through the vadose zone. Contaminants released from the surface invade the vadose zone and, before reaching the aquifer system underneath, can be altered, retarded or wholly removed by biological, chemical and physical processes in the vadose zone. Unsaturated processes control also the availability of water for agriculture, and are the driving mechanisms in slope stability, floods and other major engineering geology problems. However, in practice, the hydrology of the vadose zone is poorly known, mainly because of technical limitations in sampling and access just at one or two metres below ground. The most useful measurements of unsaturated zone conditions (moisture content via TDR and suction via tensiometers) are limited in depth to no more than a couple of metres depth and extensive monitoring over large areas is labour intensive and time consuming and are essentially local scale measurements. Other techniques are, therefore, needed. The vadose zone deeper than a couple of metres below ground surface can be mapped from the surface, at the expense of severe resolution losses, and more efficiently using borehole geophysical methods: single-borehole, borehole-to-borehole and borehole-to-surface geophysical measurements achieve a resolution sufficient for quantitative hydrologic interpretation. In most cases, the ultimate goal is the identification of hydraulic properties and parameters of the vadose zone. The dependence of the geophysical response on changes in soil moisture content, e.g., via changes in electrical resistivity or dielectric properties, is the key mechanism that permits the use of non-invasive techniques to monitor the vadose zone in time-lapse mode, i.e., via repeated measurements over time. The use of these techniques in different configurations in the shallow and deep vadose zones can provide high-resolution images of hydrogeological structures and, in some cases, a detailed assessment of dynamic processes in the subsurface environment. Both natural infiltration processes and specifically designed tracer tests can be monitored over periods of time that can last from a few hours to several years. The data from non-invasive techniques can subsequently be used to calibrate physical-mathematical models of water flow in the unsaturated zone.

Soil

Soil is a non-renewable resource, subjected to serious threats such as soil erosion, decline in organic matter, local and diffuse contamination, sealing, compaction, decline in biodiversity, salinization, floods and landslides. Soil protection and restoration require high-resolution information about soil properties. Conventional, sample-based soil property mapping is very time-consuming, cost-intensive, and the data collected are available only for discrete points in a landscape. Various soil parameters can be mapped using rapid, nearly non-destructive methods (geophysics, spectroscopy), for quasi-continuous 2D as well as 3D mapping of soil physical and hydrological properties: electromagnetic induction (EMI); GPR; magnetics, seismics, gamma ray spectrometry.

Soil is also the shallowest portion of the vadose zone. One of the key aspects of hydrological characterization of soil relies on our capability of mapping soil moisture content in its space-time variations. A leap forward was made by the introduction of time domain reflectometry (TDR) in the early 1980s, that provided a means for quick and inexpensive time-lapse monitoring of moisture content in the first couple of metres. TDR paved the way towards the general acceptance of indirect measurements of soil moisture content. Consequently, the estimation of water content space and time variations in shallow soil layers using in-situ non invasive techniques has been the focus of intensive research over the past three decades, with particular attention to techniques that measure the dielectric constant of the porous media, that is strongly affected by the presence of water in the soil pores. Among other techniques, two methods have gained popularity for the in-situ estimation of soil moisture content: surface-to-surface GPR and ERT. GPR used in surface-tosurface configuration has increasingly been proposed as a means to obtain fast and inexpensive images of soil moisture content over large areas at shallow depth. In surface-to-surface GPR measurements, the velocity of direct ground waves that travel from the transmitting antenna to the receiving antenna just below the soil surface can be used to estimate the soil water content at shallow depths over relatively large areas, at a scale of interest for hydrological models. Generally, three different survey types i.e., Wide Angle Reflection and Refraction (WARR); Common Mid Point (CMP); Single Offset (SO) methods can be used to estimate the direct ground wave velocity and infer the water content. The success of the method relies on the assumption that identifying such direct arrival is straightforward and cannot be confused with other events. This condition is not always met, e.g., in the presence of critically refracted radar waves or guided waves.

Geophysical methods have also shown some success in characterization of structural properties of the shallow soil. Such properties have direct links to hydrological responses but also impact on the ecological function of the soil.

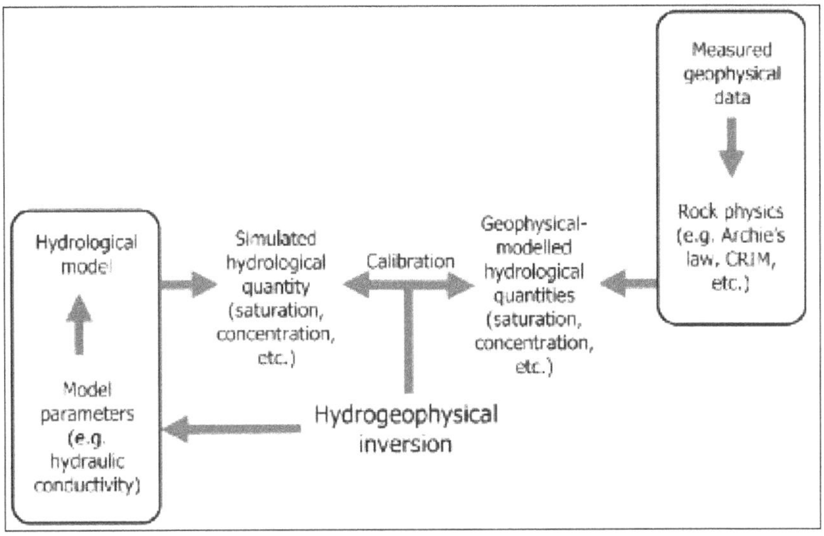

General scheme of hydrogeophysical inversion based on calibration of hydrological models on estimated hydrological quantities derived from geophysical data.

General Hydrogeophysics Approach

Of the three key aspects of the hydrological problem, it goes beyond doubt that the most important and general aspect is the understanding of fluid-dynamics. This aspect effectively is key to all hydrologically-controlled environmental problems. Irrespective of the hydrological compartment of interest, two general frameworks have been adopted to address the understanding of subsurface fluid-dynamics:

- A direct link has been sought between geophysical (e.g., electrical resistivity) and hydrological parameters (e.g., hydraulic conductivity).
- A link has been exploited between geophysical quantities and hydrological state variables (e.g., moisture content or water salinity), and identification of hydrological parameters via model calibration.

The first approach dates back to the earliest attempts to relate geophysics to hydrological parameters and in its original formulations sought essentially a link between electrical resistivity and hydraulic conductivity. This approach has had limited success, as the uncertainty associated to such link is far too wide and site-specific to be of any practical use. Some more sophisticated recent findings may lead to future developments using other geophysical techniques.

The second approach, is based on the more reliable links that can be established between geophysical quantities and hydrological variables such as water content and solute concentration, generally in the form of empirical or semiempirical relationships (constitutive relationships). Consider e.g., the classical relationships proposed by Archie for electrical conductivity and by Topp et al. And Roth et al. For the dielectric constant. Using such relationships, it is possible, albeit not always straightforward, to obtain quantitative estimates of hydrologic data to be used for the Calibration of hydrologic models, thus identifying the parameters of interest. Intuitively, the calibration of a hydrological model requires that the system is adequately stressed so that different states of the variables are explored. This condition is not guaranteed to be met under natural conditions. Consequently, it is faster and more informative to conduct controlled experiments that can be monitored in a time span of hours, days, or weeks depending on the dynamics of the system. Key to the possibility of using this approach is the capability of measuring geophysical quantities repeatedly, over time. It is interesting to note that although numerous synthetic model studies have been utilised to demonstrate quantification of hydraulic properties from timelapse geo-physical studies, very few have illustrated the approach in fieldbased studies.

Today many shallow geophysical techniques have the potential to highlight two concurring aspects of the subsurface:

- Its static aspects, i.e., the characteristics that do not change over time, principally the geology.
- Its dynamic aspects, i.e., the characteristics that do change over time, and that are inherently linked to the motion of fluids, water above all.

In this respect, the development of shallow geophysical methods is very similar to that of classical deep applications, such as reflection seismics, as utilized in the petroleum industry. More and more often, petroleum geophysics is seen as a means for understanding the nature of a site and its evolution in terms of fluid-dynamics, such as changes in fluid saturation (e.g.,, water versus oil or gas), and is being used as a key supporting technique for a site's management during petroleum production (timelapse surveying). Similarly, the future of environmental geophysics stays in its capacity to describe a site in its current state and evolution, and in the integration of this information into the calibration of hydrological, hydrogeological and geomechanical predictive models. If time-lapse measurements are made, changes in geophysical response at the same spatial location are generally linked to hydrological changes, e.g., in the vadose zone to changes in water saturation. Features that are not changing over time can be reasonably attributed to geological control: e.g., water saturation may be consistently different at different spatial locations as a consequence of lithological differences. This use of geophysical data requires:

- That the collected geophysical data have a clear, identifiable and possibly quantitative meaning in terms of environmental variables of interest, e.g., water saturation: petrophysical constitutive models are needed for this conversion; the experience of borehole geophysical logs and tdr provide the starting point for these.

- That the resolution and sensitivity of geophysical methods in space and time is fully understood, in order to assess the actual information content in the data, and prevent, e.g.,, the erroneous interpretation of artefacts.
- That hydrologic modelling be devised to be able to incorporate the non-invasive data in the most profitable and effective way, accounting for resolution, sensitivity and scale effects.

The three points above are, as of today, only partly achieved. The efforts aimed at fulfilling these requirements constitute the body of a fascinating and diversified area of research.

Opportunities in Hydrogeophysics

The rapid development of hydrogeophysical techniques over the past two decades has led to tremendous progress in a number of areas of practical and scientific interest. However, a number of opportunities and challenges remain unexplored.

Areas of New Developments

Hydrogeophysical techniques have been focused for the most part on the areas of applications described above (saturated, unsaturated zones and soil). These areas comprise a large portion of interests in hydrological sciences, but are far from completing the range of problems where data, extensive both in space and time, are needed. One of the key limitations of the most commonly used hydrogeophysical techniques is their scale of application (metres to hundreds of metres): the type of time-lapse measurements described above cannot easily exceed this limited scale. However, the majority of hydrological problems requires, in principle, monitoring and parameter assessment over the much larger catchment scale (km to thousand of km). Key hydrological questions, including the vulnerability of hydrological systems to climate changes, can only be answered in this larger setting.

Many important hydrological environments, such as shallow fresh and brackish waters, have not received so far sufficient attention in terms of hydrogeophysical characterization. One such environment is hill and mountain slopes. The understanding of hillslope processes is key to an appropriate modelling of catchment response. However, only very few examples of hydrogeophysical hillslope studies are available to date.

Hydrological science has also been developing towards the understanding of more complex systems, particularly at critical interfaces in the hydrosphere. In many areas, understanding the interaction between surface and subsurface water bodies is key to successful management of water resources. In particular, the hydrological and bioeochemical mechanisms related to nutrient transformation in the hyporheic zone is poorly understood and there is a need for new field-based methods to provide improved characterization. Very little efforts have been devoted so far to the use of hydrogeophysical techniques in this area, but this is an area of necessary development.

Improved characterization of hydrological-ecological interactions is also emerging as an area of potential development in hydrogeophysics. Robinson et al. and Kettridge et al., for example, demonstrate the value of geophysical methods for understanding spatial patterns in plant ecology, while Petersen and Al Hagrey investigate the potential of electrical methods for mapping root zone distribution. As food security, optimised management of agriculture and the preservation of ecologically vulnerable areas grow in importance, there is a clear future need for growth in hydrogeophysical developments in this area.

Another field where non-invasive characterization techniques have been used, but maybe not to their full potential, is the study of contaminated sites. All aspects of soil, unsaturated zone and saturated zone characterization are to be considered at these sites, but the presence and motion of contaminants are a further key aspect. In addition, monitoring of in-situ remediation techniques is a very promising area of time-lapse hydrogeophysical research .

The integration of geophysical data into digital soil mapping is an area of active research. However, the current techniques are deficient in terms of the understanding of relationships between mapped soil parameters and relevant soil functions, the scaling up for investigating large areas (e.g., catchments and landscapes) and the evaluation of soil degradation at such scales. In this context new strategies must be developed to derive from multi-parameter geophysical measurements the necessary information to derive unique estimates of specific soil parameters.

New/other Techniques

The core of hydrogeophysical research has been long based on few well established techniques, particularly ERT and GPR. However, novel techniques have attracted a growing interest in recent years. Spectral Induced Polarization (SIP) is an extension of the classical induced polarization that uses multi-frequency measurements, typcially in the 0.01 Hz to 1 kHz range. From a hydrological perspective, SIP (and IP) offer useful information as the measured signals are influenced by interactions along the grain boundary, rather than entirely in the pore space (as in bulk resistivity, for example). Recent evidence exists that SIP response of porous media can be linked to their structure and hydraulic properties, the presence of free-phase contaminants and microbiological activit. Successful SIP field applications have been reported but substantial technological and scientific developments are strongly needed.

Spontaneous Potential (SP) is based on the measurement of electrical potential differences present in the ground. The origin of SP has two main components: (1) the electrokinetic contribution associated with groundwater flow through the permeable soil and (2) oxidoreduction phenomena. Both can have very interesting applications in the study of water flow and contamination. However, some major progress is needed to be able to identify, case by case, the relative contribution of different SP sources to the observed field SP data.

Like SP, the measurement of gravity acceleration is not a new technique. However, the sensitivity of modern micro-gravity meters allows for accurate time-lapse measurements of gravity changes due to changes of water storage underground. The impact on large-scale hydrological monitoring can be tremendous, as microgravimetry allows for extensive and precise monitoring of mass storage in the system, and can be potentially used to calibrate full scale catchment hydrology models.

EM methods have been long utilized in the frequency and time domain, but their hydrogeophysical applications have been limited. This is mainly due to the apparent difficulty at identifying the precise support volume of the measurements and the often invoked 1D nature of EM inversion. However, EM methods have also the advantage of providing data on subsoil electrical conductivity with no need for galvanic contact. This advantage leads to the possibility of collecting data from airborne platforms over large areas in very short time. Airborne TimeDomain EM methods are strong candidates for large scale hydrogeophysical monitoring.

Surface proton magnetic resonance sounding (MRS), or nuclear magnetic resonance (NMR), measurements can be used to indirectly estimate the water content of saturated and unsaturated zones in the Earth's subsurface. MRS can be used to estimate aquifer properties, including quantity of water contained in the aquifer, porosity, and hydraulic permeability, thus being one of the most promising areas of active hydrogeophysical research. Up to now, the main limitations of MRS stand in their low signal-to-noise ratio, that make them suitable only to regions with limited anthropogenic presence.

Rock Physics

The availability of reliable constitutive models linking geophysical and hydrological quantities is key to the overall success of hydrogeophysics. A number of models exist, e.g., for electrical properties, even though many of them require site-specific parameters that need calibration, and some such parameters are not necessarily independently identified. The need exists for advanced models, accounting also for multiple links between geophysical quantities (e.g., electrical conductivity and dielectric constant) with hydrological quantities (e.g., moisture content, water salinity) having parameters with defined meaning in terms of soil/rock characteristics. These multi-parameter models, still in their infancy, are essential in order to devise quantitative means for joint hydrogeophysical data inversion. Among the tools applicable to investigate these relationships are pore-scale models.

Data Fusion and Joint Inversion

The availability of multiple geophysical data on the same subsurface structure calls for effective data integration. This can be done in many ways. One of the most promising approaches is the "Stochastic Engine" approach developed at the Lawrence Livermore Laboratory. This approach explores multiple realisations of subsurface geophysical structures (conditioned on a priori soft and hard information) as models that are consistent with the observed data. The technique, which follows a Markov chain Monte

Carlo (McMC) approach, uses geophysical forward models to test for consistency between proposed models and measured response. Although potentially computationally prohibitive for routine applications at present, the method has significant advantages over conventional deterministic approaches: (i) a wide range of data types may be integrated; (ii) the final result does not suffer from artefacts of regularisation in an inversion algorithm; (iii) estimates of model uncertainty naturally result from the model search. Within such an approach, one may also adopt a hydrological model as a means of proposing potential geophysical models. This was tested (albeit with a much simpler search algorithm) on ERT/GPR data applied to a vadose zone study by Looms et al. More sophisticated search approaches have been developed recently and we anticipate greater development and application of such approaches in the future.

In figure, we outline a basic workflow for hydrogeophysical inversion. The key elements worthy of highlight are: (1) the hydrological conceptual model is key to geophysical survey design; (2) the rock physics (constitutive relationships) are likely to be uncertain and this must be recognised in model proposals; (3) there is likely to be a range of possible hydrological models that are consistent with the observed data; (4) multiple data sources should be used to help propose models and test outcomes.

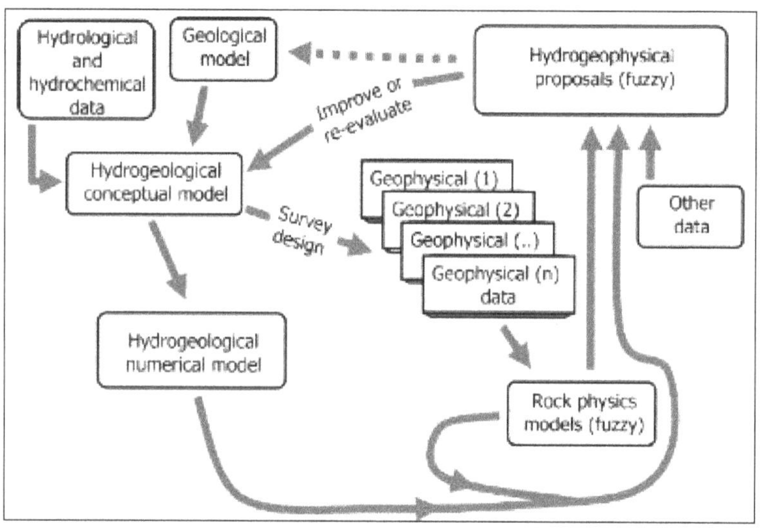

Proposed basic workflow of hydrogeophysical inversion based on calibration of hydrological models directly on geophysical data.

Challenges and Limitations

The tremendous potential of hydrogeophysics faces also a number of limitations; challenges for the future are to overcome or at least circumvent some of these limitations. The most notable ones are:

- Geophysical data are limited in terms of resolution. This is true even in cross-hole configuration. The key consequence of this is that not the entire geophysical "image" has the same degree of accuracy and practically all inversion methods require some sort of a-priori information (e.g., smoothness) to ensure convergence. Interpreting (also, quantitatively) these hydrogeophysical images with no account for resolution limitations can lead to serious problems. The most notable one is the mass balance errors very often present in tracer tests monitored e.g., via ERT both in the saturated and the unsaturated zones;

- Scale issues: most hydrogeophysical work has been conducted to date at the small scale. But hydrology needs large scale evaluation. For this reason some techniques that hold the promise to develop into large scale time-lapse monitoring tools should be considered as top priority research areas;
- Inversion and joint inversion is still computationally very expensive, especially in 3D and time-lapse. This, together with the acquisition time in the field, still limits our capability of fully exploiting the conceptual potential of hydrogeophysics.

References

- Reynolds, John (2011). Introduction to Applied & Environmental Geophysics. Wiley-Blackwell. ISBN 978-0-471-48535-3
- Environmental-geophysics-Techniques-advantages-and-limitations-298874887, Retrieved 08 January, 2020
- Fathi, Arash; Poursartip, Babak; Kallivokas, Loukas (2015). "Time-domain hybrid formulations for wave simulations in three-dimensional PML-truncated heterogeneous media". International Journal for Numerical Methods in Engineering. 101 (3): 165–198. Bibcode:2015IJNME.101..165F. doi:10.1002/nme.4780
- Hydrogeophysics-Opportunities-and-Challenges-228648418: researchgate.net, Retrieved 13 March, 2020
- Rebetsky, Yu. L. (2009). "Modern problems of tectonophysics". Izvestiya, Physics of the Solid Earth. 45 (11): 933–937. Bibcode:2009IzPSE..45..933R. doi:10.1134/S1069351309110019

Geodynamics 3

- Geodynamics of Terrestrial Exoplanets
- Geothermal Gradient
- Mineral Physics
- Dynamic Topography

The sub-field of geophysics that deals with the dynamics and geologic phenomena of the Earth such as seafloor spreading, mountain building, volcanoes, earthquakes and faulting, is termed as geodynamics. This chapter has been carefully written to provide an extensive understanding of the geodynamics.

Geodynamics is a part of geophysics dealing with the different dynamics in the Earth. Geodynamics studies, in particular, how the convection in the layer of the Earth called the mantle can lead to plate tectonics. A broad, multidisciplinary branch of geophysics that investigates the causes and consequences of the forces acting on and inside the solid Earth and the rock movements and deformations they produce. The term geodynamics originates from two root words geo-, relating to the Earth, and dynamics, the branch of physics that involves the study of forces acting on bodies and the changes in motion they produce. As the name therefore suggests, the focus of geodynamics is on the forces that act on the solid Earth and the motions and deformation they produce.

Deformation of Rocks

Within the Earth rocks are continually being subjected to forces that tend to bend them, twist them, or fracture them. When rocks bend, twist or fracture we say that they deform (change shape or size). The forces that cause deformation of rock are referred to as stresses (Force/unit area). So, to understand rock deformation we must first explore these forces or stresses.

Stress and Strain

Stress is a force applied over an area. One type of stress that we are all used to is a uniform stress, called pressure. A uniform stress is a stress wherein the forces act equally from all directions. In the Earth the pressure due to the weight of overlying rocks is a uniform stress, and is sometimes referred to as confining stress.

If stress is not equal from all directions then we say that the stress is a differential stress. Three kinds of differential stress occur:

- Tensional stress (or extensional stress), which stretches rock.
- Compressional stress, which squeezes rock.
- Shear stress, which result in slippage and translation.

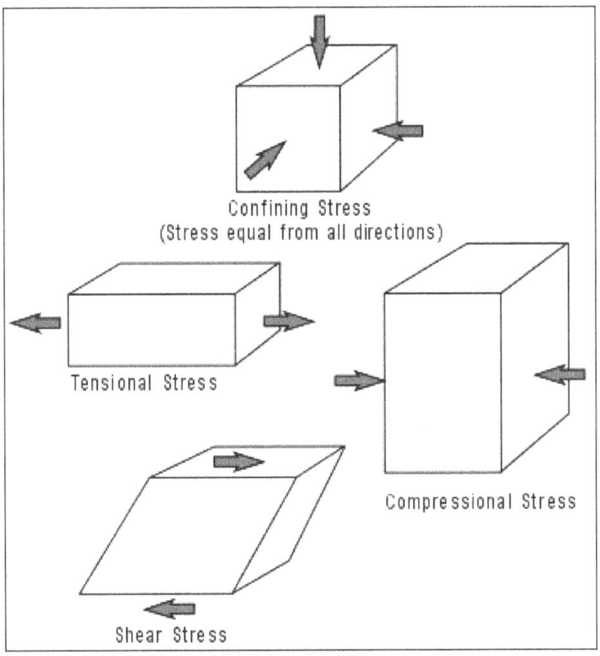

When rocks deform they are said to strain. A strain is a change in size, shape, or volume of a material.

Stages of Deformation

When a rock is subjected to increasing stress it passes through 3 successive stages of deformation:

- Elastic Deformation: Wherein the strain is reversible.
- Ductile Deformation: Wherein the strain is irreversible.
- Fracture: Irreversible strain wherein the material breaks.

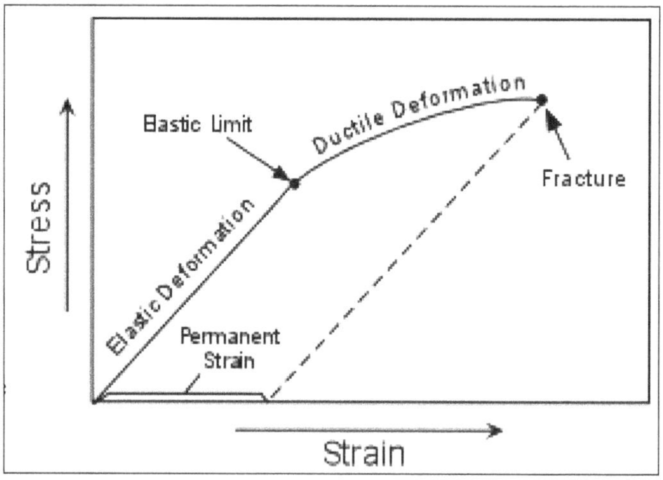

We can divide materials into two classes that depend on their relative behavior under stress:

- Brittle materials have a small or large region of elastic behavior but only a small region of ductile behavior before they fracture.
- Ductile materials have a small region of elastic behavior and a large region of ductile behavior before they fracture.

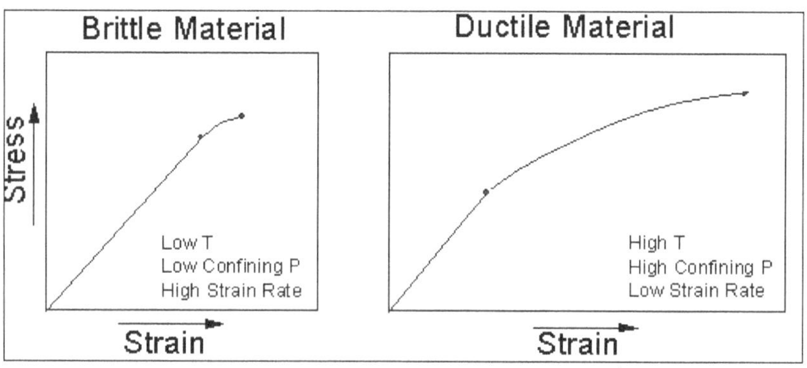

How a material behaves will depend on several factors. Among them are:

- Temperature: At high temperature molecules and their bonds can stretch and move, thus materials will behave in more ductile manner. At low temperature, materials are brittle.
- Confining Pressure: At high confining pressure materials are less likely to fracture because the pressure of the surroundings tends to hinder the formation of fractures. At low confining stress, material will be brittle and tend to fracture sooner.
- Strain rate: At high strain rates material tends to fracture. At low strain rates more time is available for individual atoms to move and therefore ductile behavior is favored.

- Composition: Some minerals, like quartz, olivine, and feldspars are very brittle. Others, like clay minerals, micas, and calcite are more ductile. This is due to the chemical bond types that hold them together. Thus, the mineralogical composition of the rock will be a factor in determining the deformational behavior of the rock. Another aspect is presence or absence of water. Water appears to weaken the chemical bonds and forms films around mineral grains along which slippage can take place. Thus wet rock tends to behave in ductile manner, while dry rocks tend to behave in brittle manner.

Brittle-ductile Properties of the Lithosphere

We all know that rocks near the surface of the Earth behave in a brittle manner. Crustal rocks are composed of minerals like quartz and feldspar which have high strength, particularly at low pressure and temperature. As we go deeper in the Earth the strength of these rocks initially increases. At a depth of about 15 km we reach a point called the brittle-ductile transition zone. Below this point rock strength decreases because fractures become closed and the temperature is higher, making the rocks behave in a ductile manner. At the base of the crust the rock type changes to peridotite which is rich in olivine. Olivine is stronger than the minerals that make up most crustal rocks, so the upper part of the mantle is again strong. But, just as in the crust, increasing temperature eventually predominates and at a depth of about 40 km the brittle-ductile transition zone in the mantle occurs. Below this point rocks behave in an increasingly ductile manner.

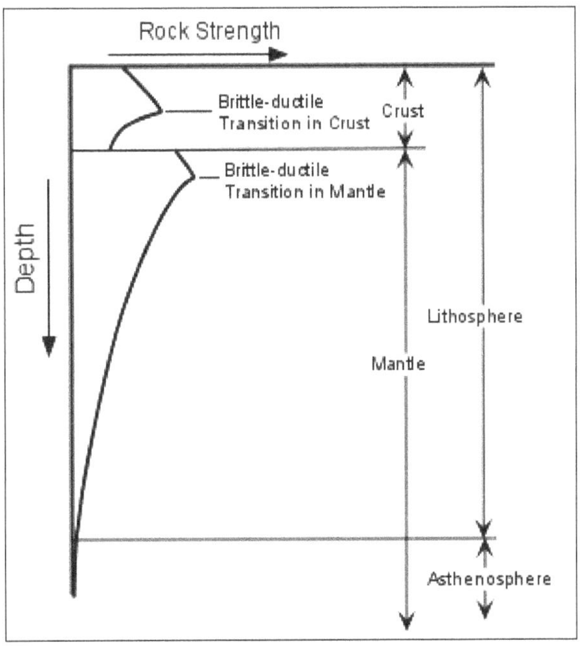

Only in a few cases does deformation of rocks occur at a rate that is observable on human time scales. Abrupt deformation along faults, usually associated with earthquakes caused by the fracture of rocks occurs on a time scale of minutes or seconds. Gradual deformation along faults or in areas of uplift or subsidence can be measured over periods of months to years with sensitive measuring instruments.

Evidence of Former Deformation

Evidence of deformation that has occurred in the past is very evident in crustal rocks. For example, sedimentary strata and lava flows generally follow the law of original horizontality. Thus, when we see such strata inclined instead of horizontal, evidence of an episode of deformation. In order to uniquely define the orientation of a planar feature we first need to define two terms - strike and dip.

For an inclined plane the strike is the compass direction of any horizontal line on the plane. The dip is the angle between a horizontal plane and the inclined plane, measured perpendicular to the direction of strike.

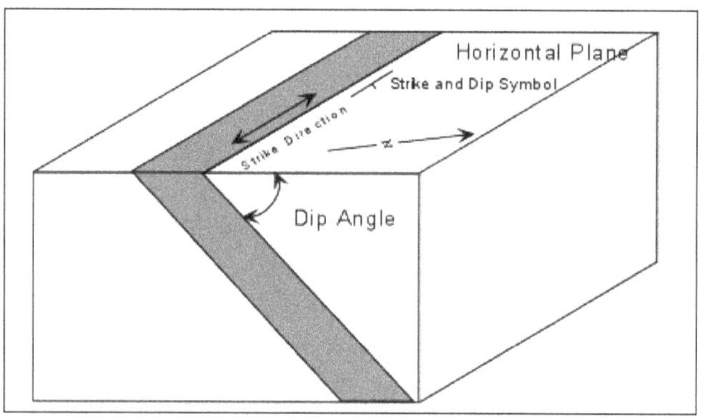

In recording strike and dip measurements on a geologic map, a symbol is used that has a long line oriented parallel to the compass direction of the strike. A short tick mark is placed in the center of the line on the side to which the inclined plane dips, and the angle of dip is recorded next to the strike and dip symbol as shown above. For beds with a 90° dip (vertical) the short line crosses the strike line, and for beds with no dip (horizontal) a circle with a cross inside is used as shown below.

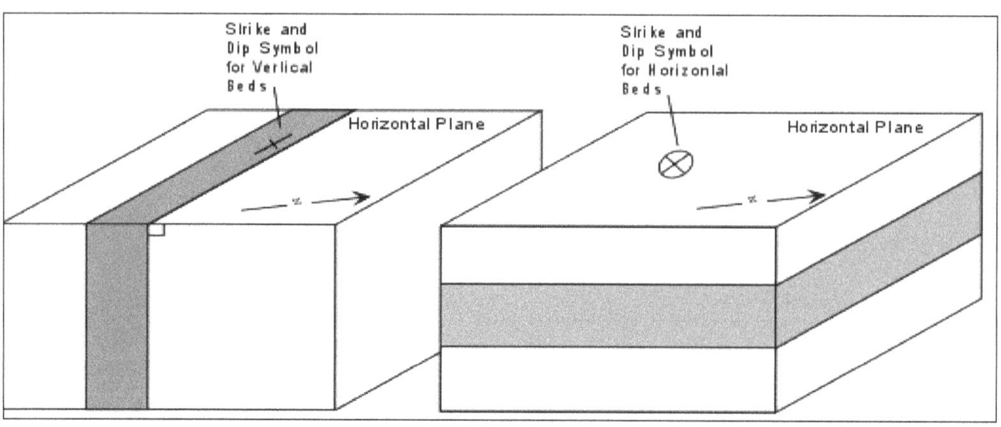

Fracture of Brittle Rocks

Faults occur when brittle rocks fracture and there is an offset along the fracture. When the offset is small, the displacement can be easily measured, but sometimes the displacement is so large that it is difficult to measure.

Types of Faults

Faults can be divided into several different types depending on the direction of relative displacement. Since faults are planar features, the concept of strike and dip also applies, and thus the strike and dip of a fault plane can be measured. One division of faults is between dipslip faults, where the displacement is measured along the dip direction of the fault, and strikeslip faults where the displacement is horizontal, parallel to the strike of the fault.

Dip Slip Faults

Dip slip faults are faults that have an inclined fault plane and along which the relative displacement or offset has occurred along the dip direction. Note that in looking at the displacement on any fault we don't know which side actually moved or if both sides moved, all we can determine is the relative sense of motion.

For any inclined fault plane we define the block above the fault as the hanging wall block and the block below the fault as the footwall block.

Normal Faults

Normal Faults are faults that result from horizontal tensional stresses in brittle rocks and where the hanging-wall block has moved down relative to the footwall block.

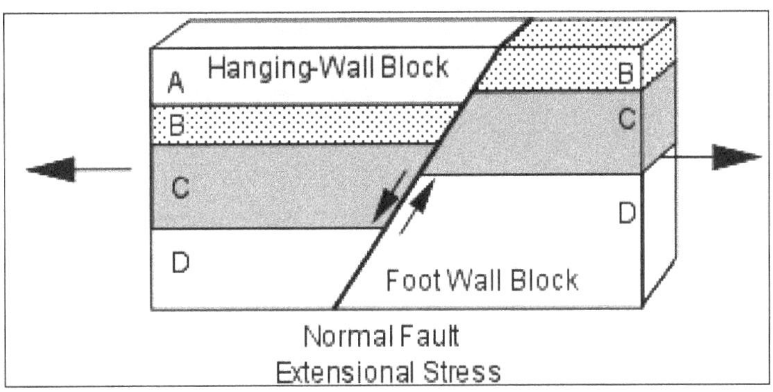

Horsts and Grabens

Due to the tensional stress responsible for normal faults, they often occur in a series, with adjacent faults dipping in opposite directions. In such a case the down-dropped blocks form grabens and the uplifted blocks form horsts. In areas where tensional stress has recently affected the crust, the grabens may form rift valleys and the uplifted horst blocks may form linear mountain ranges. The East African Rift Valley is an example of an area where continental extension has created such a rift. The basin and range province of the western U.S. is also an area that has recently undergone crustal extension. In the basin and range, the basins are elongated grabens that now form valleys, and the ranges are uplifted horst blocks.

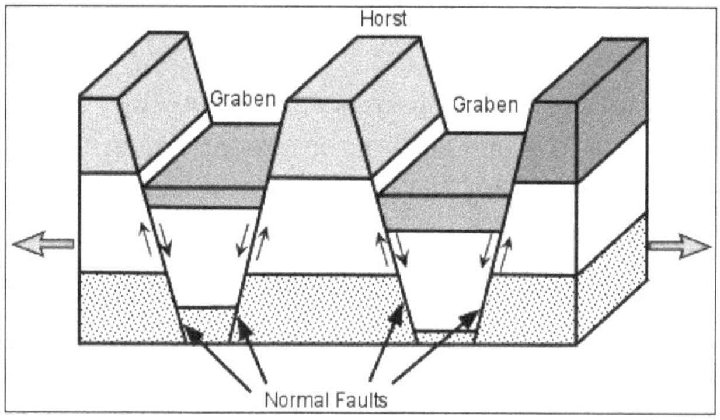

Half-Grabens

A normal fault that has a curved fault plane with the dip decreasing with depth can cause the down-dropped block to rotate. In such a case a half-graben is produced, called such because it is bounded by only one fault instead of the two that form a normal graben.

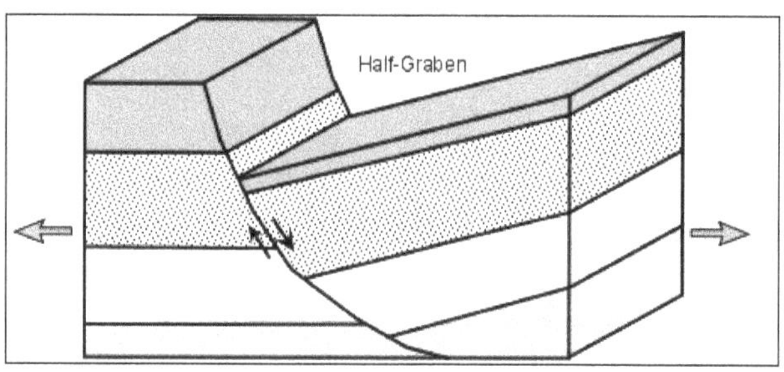

Reverse Faults

Reverse Faults are faults that result from horizontal compressional stresses in brittle rocks, where the hanging-wall block has moved up relative the footwall block.

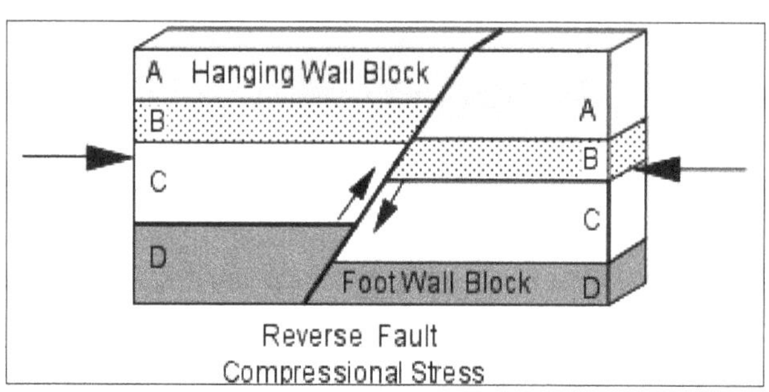

Thrust Fault

A Thrust Fault is a special case of a reverse fault where the dip of the fault is less than 15° . Thrust faults can have considerable displacement, measuring hundreds of kilometers, and can result in older strata overlying younger strata.

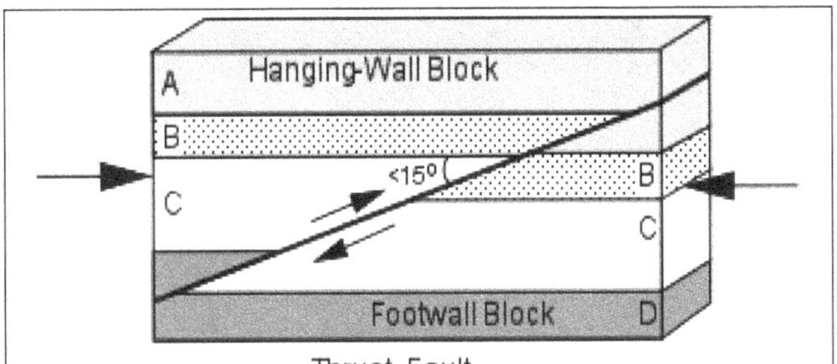

Thrust Fault

Strike Slip Faults

Strike Slip Faults are faults where the relative motion on the fault has taken place along a horizontal direction. Such faults result from shear stresses acting in the crust. Strike slip faults can be of two varieties, depending on the sense of displacement. To an observer standing on one side of the fault and looking across the fault, if the block on the other side has moved to the left, we say that the fault is a left-lateral strike-slip fault. If the block on the other side has moved to the right, we say that the fault is a right-lateral strike-slip fault. The famous San Andreas Fault in California is an example of a right-lateral strike-slip fault. Displacements on the San Andreas fault are estimated at over 600 km.

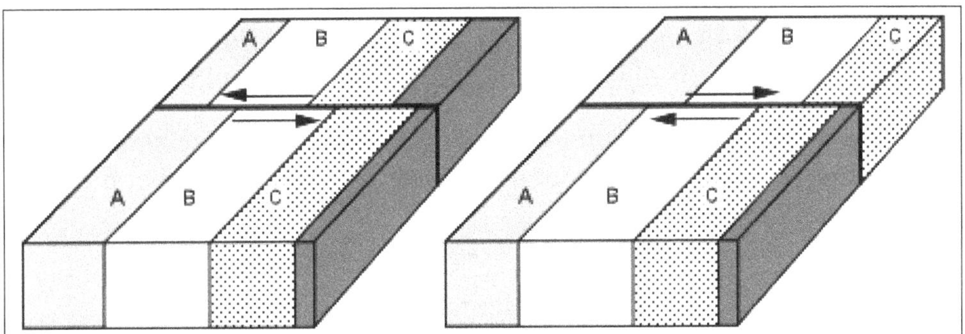

Left-Lateral Strike-Slip Fault   Right-Lateral Strike-Slip Fault

Transform-faults

These are a special class of strike-slip faults. These are plate boundaries along which two plates slide past one another in a horizontal manner. The most common type of transform faults occur where oceanic ridges are offset. Note that the transform fault only occurs between the two segments of the ridge. Outside of this area there is no relative movement because blocks are moving in the same direction. These areas are called fracture zones. The San Andreas fault in California is also a transform fault.

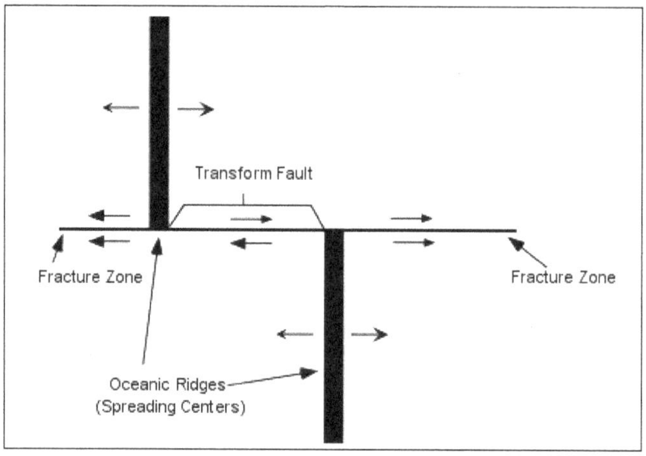

Evidence of Movement on Faults

- Slikensides are scratch marks that are left on the fault plane as one block moves relative to the other. Slickensides can be used to determine the direction and sense of motion on a fault.
- Fault Breccias are crumbled up rocks consisting of angular fragments that were formed as a result of grinding and crushing movement along a fault.

Folding of Ductile Rocks

When rocks deform in a ductile manner, instead of fracturing to form faults, they may bend or fold, and the resulting structures are called folds. Folds result from compressional stresses acting over considerable time. Because the strain rate is low, rocks that we normally consider brittle can behave in a ductile manner resulting in such folds

We recognize several different kinds of folds. Monoclines are the simplest types of folds. Monoclines occur when horizontal strata are bent upward so that the two limbs of the fold are still horizontal.

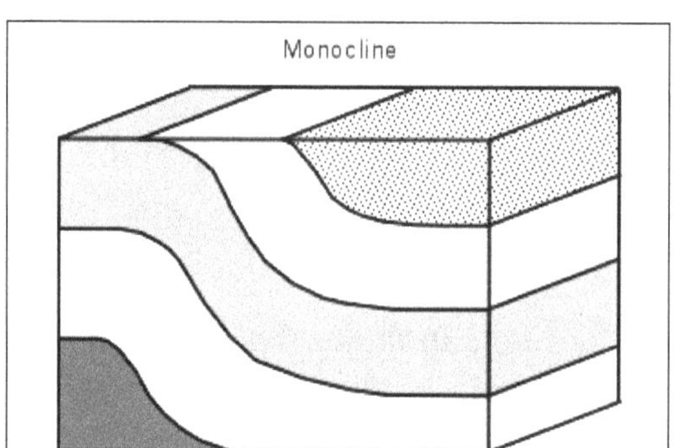

Anticlines are folds where the originally horizontal strata has been folded upward, and the two limbs of the fold dip away from the hinge of the fold.

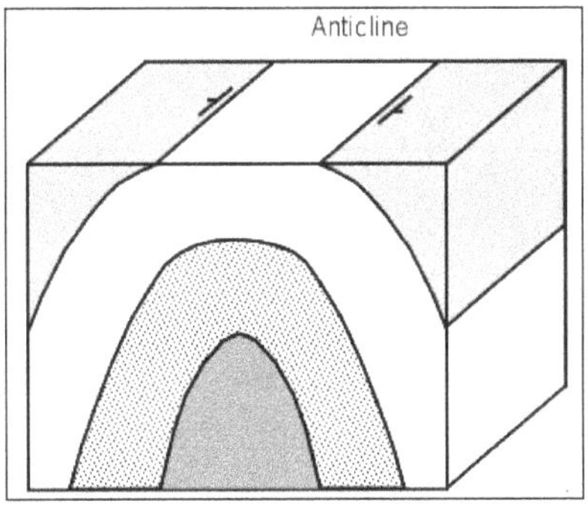

Synclines are folds where the originally horizontal strata have been folded downward, and the two limbs of the fold dip inward toward the hinge of the fold. Synclines and anticlines usually occur together such that the limb of a syncline is also the limb of an anticline.

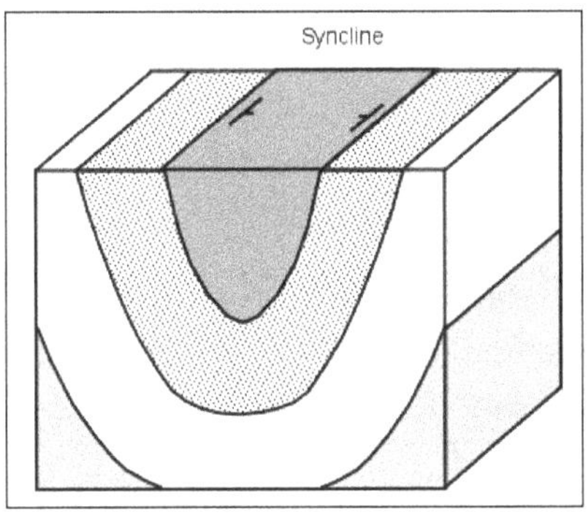

Geometry of Folds

Folds are described by their form and orientation. The sides of a fold are called limbs. The limbs intersect at the tightest part of the fold, called the hinge. A line connecting all points on the hinge is called the fold axis. In the diagrams above, the fold axes are horizontal, but if the fold axis is not horizontal the fold is called a plunging fold and the angle that the fold axis makes with a horizontal line is called the plunge of the fold. An imaginary plane that includes the fold axis and divides the fold as symmetrically as possible is called the axial plane of the fold.

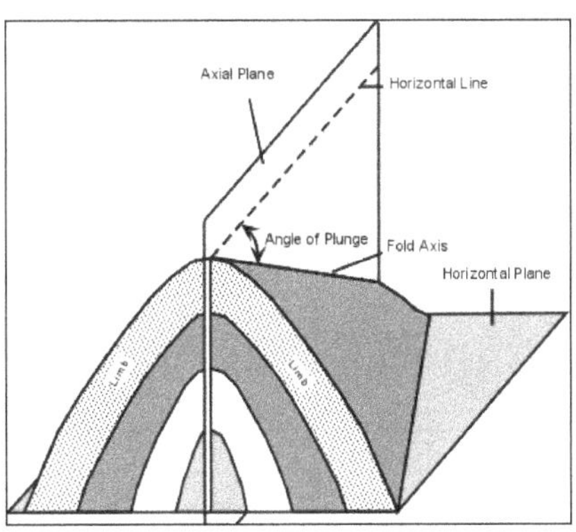

Note that if a plunging fold intersects a horizontal surface, we will see the pattern of the fold on the surface.

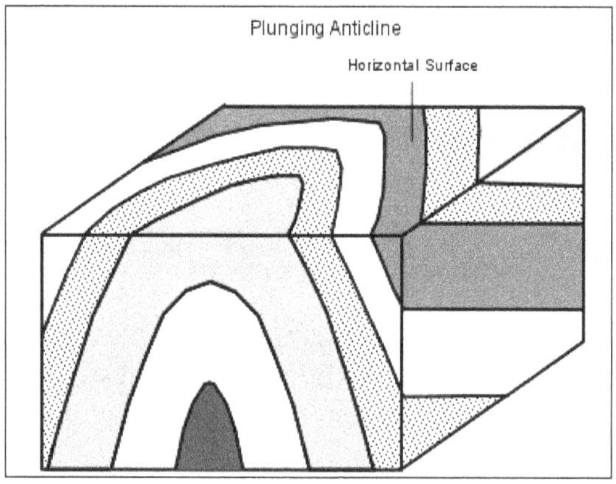

Classification of Folds

Folds can be classified based on their appearance:

- If the two limbs of the fold dip away from the axis with the same angle, the fold is said to be a symmetrical fold.

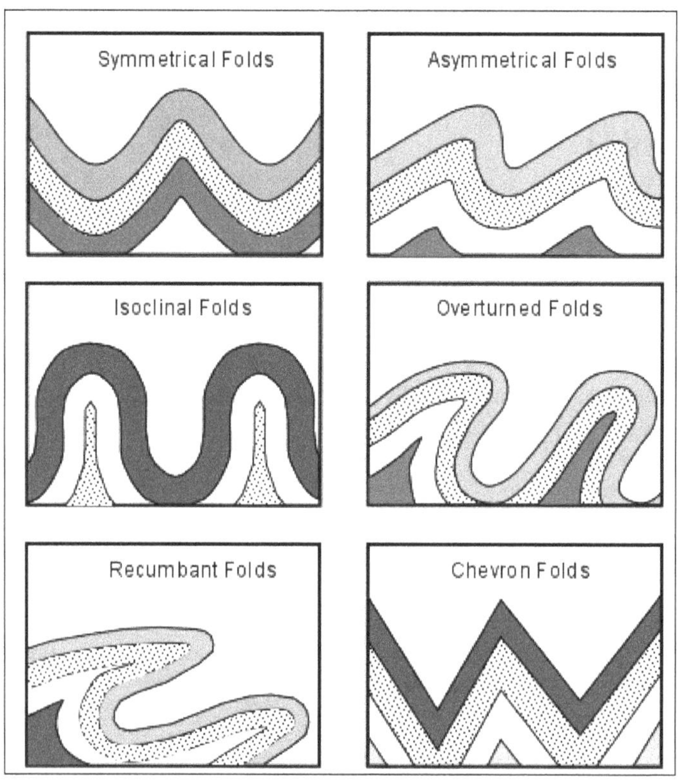

- If the limbs dip at different angles, the folds are said to be asymmetrical folds.

- If the compressional stresses that cause the folding are intense, the fold can close up and have limbs that are parallel to each other. Such a fold is called an isoclinal fold (iso means same, and cline means angle, so isoclinal means the limbs have the same angle). Note the isoclinal fold depicted in the diagram below is also a symmetrical fold.
- If the folding is so intense that the strata on one limb of the fold becomes nearly upside down, the fold is called an overturned fold.
- An overturned fold with an axial plane that is nearly horizontal is called a recumbant fold.
- A fold that has no curvature in its hinge and straight-sided limbs that form a zigzag pattern is called a chevron fold.

Relationship between Folding and Faulting

Because different rocks behave differently under stress, we expect that some rocks when subjected to the same stress will fracture or fault, while others will fold. When such contrasting rocks occur in the same area, such as ductile rocks overlying brittle rocks, the brittle rocks may fault and the ductile rocks may bend or fold over the fault.

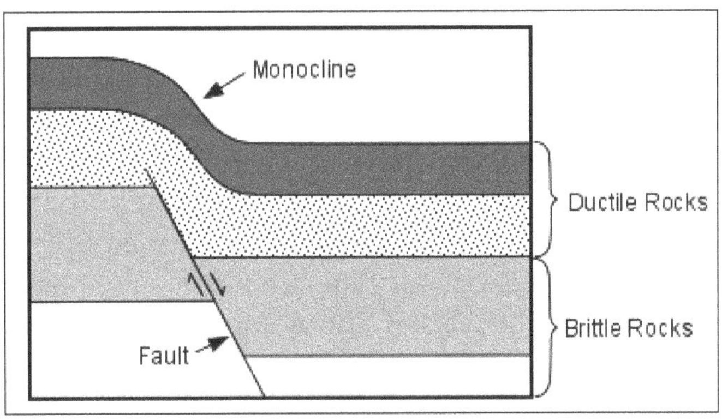

Also since even ductile rocks can eventually fracture under high stress, rocks may fold up to a certain point then fracture to form a fault.

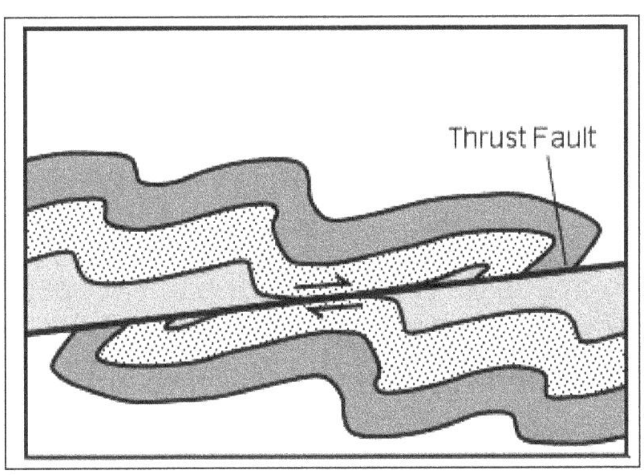

Folds and Topography

Since different rocks have different resistance to erosion and weathering, erosion of folded areas can lead to a topography that reflects the folding. Resistant strata would form ridges that have the same form as the folds, while less resistant strata will form valleys.

Mountain Ranges

One of the most spectacular results of deformation acting within the crust of the Earth is the formation of mountain ranges. Mountains originate by three processes, two of which are directly related to deformation

Thus, there are three types of mountains:

- Fault Block Mountains: As the name implies, fault block mountains originate by faulting. As discussed previously, both normal and reverse faults can cause the uplift of blocks of crustal rocks. The Sierra Nevada mountains of California, and the mountains in the Basin and Range province of the western U.S., as discussed previously, were formed by faulting processes and are thus fault block mountains.
- Fold and Thrust Mountains: Large compressional stresses can be generated in the crust by tectonic forces that cause continental crustal areas to collide. When this occurs the rocks between the two continental blocks become folded and faulted under compressional stresses and are pushed upward to form fold and thrust mountains. The Himalayan Mountains (currently the highest on Earth) are mountains of this type and were formed as a result of the Indian Plate colliding with the Eurasian plate. Similarly the Appalachian Mountains of North America and the Alps of Europe were formed by such processes.
- Volcanic Mountains: The third type of mountains, volcanic mountains, are not formed by deformational processes, but instead by the outpouring of magma onto the surface of the Earth. The Cascade Mountains of the western U.S., and of course the mountains of the Hawaiian Islands and Iceland are volcanic mountains.

Geodynamics of Terrestrial Exoplanets

The discovery of extrasolar Earth-sized planets has encouraged research into their potential for habitability. One of the generally agreed requirements for a life-sustaining planet is a mobile, fractured lithosphere cyclically recycled into a vigorously convecting mantle, in a process commonly known as plate tectonics. Plate tectonics provide a means of geochemical regulation of atmospheric particulates, as well as removal of carbon from the atmosphere. This prevents a "runaway greenhouse" effect that can result in inhospitable surface temperatures and vaporization of liquid surface water. Planetary scientists have not reached a consensus on whether Earth-like exoplanets have plate tectonics, but it is widely thought that the likelihood of plate tectonics on an Earth-like exoplanet is a function of planetary radius, initial temperature upon coalescence, insolation, and presence or absence of liquid-phase surface water.

Artistic sketch of Kepler-22b, a recently discovered exoplanet with comparable mass (within 10 Earth masses) of the planet Earth.

Potential Exoplanet Geodynamic Regimes

In order to characterize the geodynamic regime of an Earth-like exoplanet, the basic assumption is made that such a planet is Earth-like or "rocky". This implies a three-layer stratigraphy of (from center to surface) a partially molten iron core, a silicate mantle that convects over geologic timescales, and a relatively cold, brittle silicate lithosphere. Within these parameters, the geodynamic regime at a given time point in the planet's history is likely to fall within one of three categories.

Plate Tectonics

The mantle of a planet with plate tectonics has driving forces that exceed the yield strength of the brittle lithosphere, causing the lithosphere to fracture into plates that move relative to each other. A critical element of the plate tectonic system is these lithospheric plates become negatively buoyant at some point in their evolution, sinking into the mantle. The surface mass deficit is balanced by new plate being formed elsewhere through upwelling mantle plumes. Plate tectonics is an efficient method of heat transfer from the interior of the planet to the surface. Earth is the only planet plate tectonics is known to occur on, although evidence has been presented for Jupiter's moon Europa undergoing a form of plate tectonics analogous to Earth's.

Stagnant Lid

A stagnant lid regime occurs when mantle driving forces do not exceed the lithospheric yield strength, resulting in a single, continuous rigid plate overlying the mantle. Stagnant lids only develop when the viscosity contrast between the surface and planetary interior exceeds about four orders of magnitude.

Episodic Tectonics

Episodic tectonics is a general term for a geodynamic regime that possesses aspects of both plate tectonics and stagnant lid dynamics. Planets with episodic tectonic regimes will have immobile surface lids for geologically long spans of time, until a shift in equilibrium conditions is precipitated by either weakening lithosphere or increasing mantle driving forces. When this occurs, the shift to plate tectonics is usually catastrophic in nature and can involve resurfacing of the entire planet. After such a resurfacing event (or period of resurfacing events), stagnant lid equilibrium conditions are regained, resulting in a quiescent, immobile lid.

Methods of Predicting Exoplanet Geodynamic Regimes

Exoplanets have been directly observed and remotely sensed, but due to their great distance and proximity to obscuring energy sources (the stars they orbit), there is little concrete knowledge of their composition and geodynamic regime. Therefore, the majority of information and conjectures made about them come from alternative sources.

Solar System Analogues

All the rocky planets in the solar system except Earth are generally believed to be in the stagnant lid geodynamic regime. Mars and particularly Venus have evidence of prior resurfacing events, but appear to be tectonically quiescent today. Geodynamic inferences about solar system planets have been extrapolated to exoplanets in order to constrain what kind of geodynamic regimes can be expected given a set of physical criterion such as planetary radius, presence of surface water, and insolation. In particular, the planet Venus has been intensely studied due to its general physical similarities to Earth yet completely different geodynamic regime. Proposed explanations include a lack of surface water, the lack of a magnetic geodynamo, or large-scale evacuation of interior heat shortly after planetary coalescence.

Another source of insight within our solar system is the history of the planet Earth, which may have had several episodes of stagnant lid geodynamics during its history. These stagnant-lid periods were not necessarily planet-wide; when supercontinents such as Gondwanaland existed, their presence may have shut off plate motion over large expanses of the Earth's surface until mantle heat buildup underneath the superplate was sufficient to break them apart.

Observation of Exoplanets

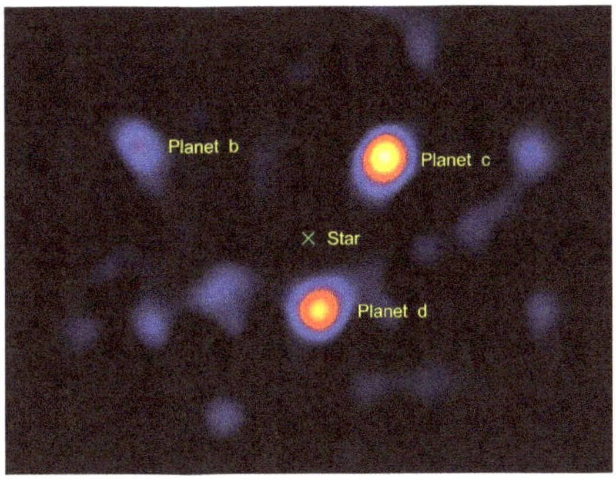

Three identified exoplanets around the roughly sun-sized star HR8799, imaged through a vector vortex coronagraph on a 1.5m section of the Hale telescope.

Indirect and direct observation methods such as radial velocity and coronagraphs can give envelope estimates of exoplanet parameters such as mass, planetary radius, and orbital radius/eccentricity. Since distance from the host star and planetary size are generally believed to influence exoplanet geodynamic regime, inferences can be drawn from such information. For example, an exoplanet close enough to its host star to be tidally locked may have drastically different "dark" and "light" side temperatures and correspondingly bipolar geodynamic regimes. Spectroscopy has been used to characterize extrasolar gas giants, but has not yet been used on rocky exoplanets. However, numerical modeling has demonstrated that spectroscopy could detect atmospheric sulfur dioxide levels as low as 1 ppm; presence of sulfur dioxide at this concentration may be indicative of a planet without surface water with volcanism 1500–80,000 times higher than Earth.

Numerical Modeling

Since real data on exoplanets is currently limited, a large amount of the dialogue regarding rocky exoplanet tectonics has been driven by the results of numerical modeling studies. In such models, different planetary physical parameters are manipulated (i.e. mantle viscosity, core-mantle boundary temperature, insolation, "wetness" or hydration of subducting lithosphere) and the resultant impact on the geodynamic regime is reported. Due to computational limitations the large amount of variables that control planet geodynamics in real life cannot be accounted for; models therefore ignore certain parameters believed to be less important and emphasize others to try to isolate disproportionately important driving factors.

Scaling Parameters

Bar chart showing the size distribution of observed Kepler planet candidates (terrestrial exoplanets in the habitable zone of their host star). Data set is 2,740 planets orbiting

2,036 stars. The Earth-size and Super Earth-size (leftmost) columns represent potential terrestrial exoplanets.

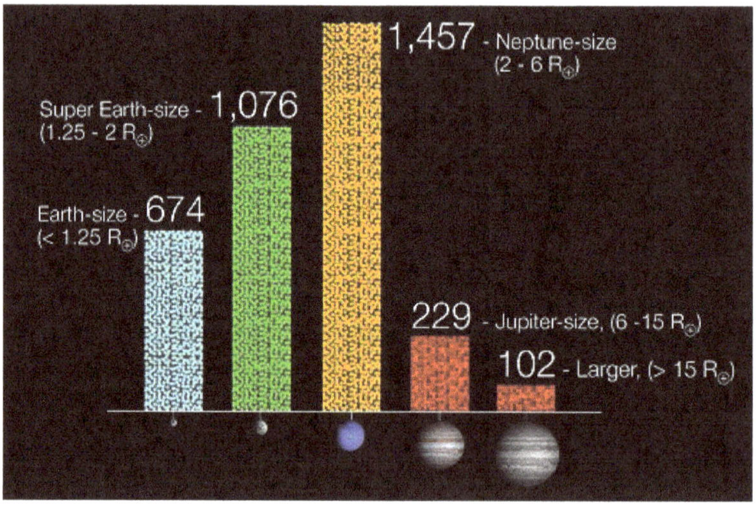

Early models of rocky exoplanets scaled different factors (namely mantle viscosity, lithospheric yield strength, and planetary size) up and down to predict the geodynamic regime of an exoplanet with given parameters. Two scaling studies of exoplanet size published in 2007 came to fundamentally different conclusions: O'Neill and Lenardic showed that a planet of 1.1 Earth mass would have Earth-like lithospheric yield stress but reduced mantle driving stresses, resulting in a stagnant lid regime. Conversely, Valencia et al. concluded the increase in mantle velocity (driving force) is large compared to the gravitationally-forced increase of plate viscosity as planets increase beyond one Earth mass, increasing the likelihood of plate tectonics with planet size.

Viscoelastic-plastic Rheology

Most models simulate lithospheric plates with a viscoelastic-plastic rheology. In this simulation, plates deform viscoelastically up to a threshold level of stress, at which point they deform in a plastic manner. The lithospheric yield stress is a function of pressure, stress, composition, but temperature has a disproportionate effect on it. Therefore, changes to the lithospheric temperature, whether from external sources (insolation) or internal (mantle heating) will increase or decrease the likelihood of plate tectonics in viscoelastic-plastic models. Models with different modes of mantle heating (heat originating from the core-mantle boundary versus in-situ mantle heating) can produce dramatically different geodynamic regimes.

Time-dependent versus Quasi-steady States

For computational purposes, early exoplanet mantle convection models assumed the planet was in a quasi-steady state, that is, the heat input from the core-mantle boundary or internal mantle heating remained constant throughout the model run. Later studies such as that of Noack and Breuer show that this assumption may have important implications, resulting in a gradual increase of the temperature differential between the core and mantle. A planet modeled with realistic decrease of internal heating throughout time had a lower likelihood of entering a plate tectonic regime compared to the quasi-steady state model.

Damage Theory

A flaw of viscoelastic-plastic models of exoplanet geodynamics is in order for plate tectonics to be initiated, unrealistically low yield stress values are required. Additionally, plates in viscoelastic-plastic models have no deformation memory, i.e. as soon as the stress on a lithospheric plate drops below its yield stress it returns to its pre-deformation strength. This stands in contrast to Earth-based observations, which show that plates preferentially break along preexisting areas of deformation.

Damage theory attempts to address this model flaw by simulating voids created in areas of strain, representing the mechanical pulverization of coarse grains of rock into finer grains. In such models, damage is balanced by "healing", or the temperature and pressure-driven dynamic recrystallization of smaller grains into larger ones. If the reduction of grain size (damage) is intensely localized in a stagnant lid, an incipient crack in the mantle can turn into a full-blown rift, initiating plate tectonics. Conversely, a high surface temperature will have more efficient lithospheric healing, which is another potential explanation for why Venus has a stagnant lid and Earth does not.

Factors for Earth-like Exoplanet Geodynamic Regimes

Initial Temperature

For rocky exoplanets larger than Earth, the initial interior temperature after planetary convalescence may be an important controlling factor of surface motion. Noack and Breuer demonstrated that a core-mantle boundary initial temperature of 6100 K would likely form a stagnant lid, while a planet of the same dimensions with an initial core-mantle boundary 2000 K hotter will likely eventually evolve plate tectonics. This effect is diminished on planets smaller than Earth, because their smaller planetary interiors efficiently redistribute heat, reducing core-mantle heat gradients that drive mantle convection.

Insolation

External sources of planetary heat (namely, radiation from a planet's host star) can have drastic effects on geodynamic regime. With all other variables held constant, an Earth-sized exoplanet with a surface temperature of 273 K will evolve over its geological lifetime from a plate tectonic regime, to episodic periods of plate tectonics interspersed with stagnant lid geodynamics, to a terminal stagnant lid phase as interior heat is exhausted. Meanwhile, a "hot" planet (759 K surface temperature) under the same initial conditions will have an amorphous surface (due to lithospheric yield stress being constantly exceeded) to a stagnant lid as interior heat is exhausted, with no plate tectonics observed.

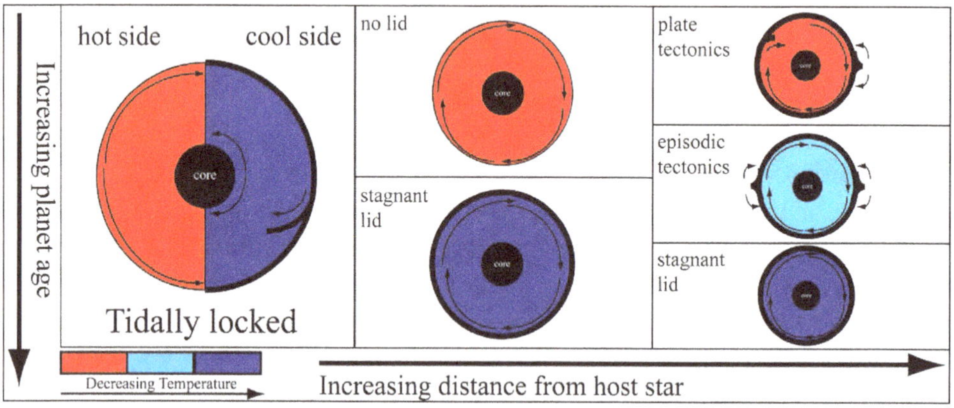

Conceptual plot of the effect of distance from a host star vs. planetary age on terrestrial exoplanet geodynamics. Example planets not drawn to scale.

Planets closer than 0.5 astronomical units from their star are likely to be tidally locked; these planets are expected to have drastically different temperature regimes on their "day" and "night" sides. When this scenario is modeled, the day side displays mobile lid convection with diffuse surface deformation flowing toward the night side, while the night side has a plate tectonic regime of downwelling plates and a deep mantle return flow in the direction of the night side. A temperature contrast of 400 K between day and night sides is required to create such a stable system.

Presence of Surface Water

While early modeling studies emphasized the size of a given exoplanet as a critical factor of geodynamic regime, later studies showed that the influence of size may be small to the point of irrelevance compared to the presence of surface water. For plate tectonics to be a sustained, rather than episodic process, the friction coefficient at the upper boundary layer (the mantle-lithosphere interface) must be below a critical value; while some models arrive at a critically low friction coefficient via increased upper boundary layer temperature (and subsequent decreased viscosity), Korenaga demonstrates high pore fluid content can lower the coefficient of friction below the critical value as well.

Implications of Exoplanet Geodynamic Regime

A planet in a stagnant lid regime has a much lower likelihood of being habitable than one with active surface recycling. The outgassing of mantle-derived carbon and sulfur that occurs along plate margins is critical for producing and maintaining an atmosphere, which insulates a planet from solar radiation and wind. The same atmosphere also regulates surface temperature, providing a clement condition for biological activity. It is for these reasons the search for exoplanets will be steered largely towards finding ones with a plate tectonic geodynamic regime, since they are better candidates for human habitation.

Geothermal Gradient

Geothermal gradient is the rate of increasing temperature with respect to increasing depth in the Earth's interior. Away from tectonic plate boundaries, it is about 25–30 °C/km (72-87 °F/mi) of depth near the surface in most of the world. Strictly speaking, geo-thermal necessarily refers to the Earth but the concept may be applied to other planets.

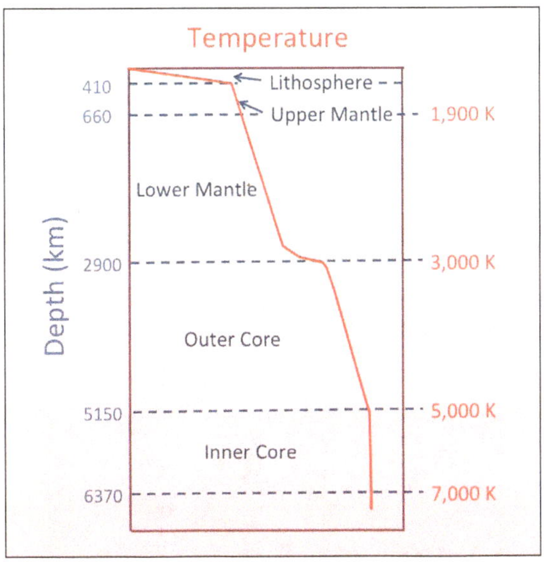

Temperature profile of the inner Earth.

The Earth's internal heat comes from a combination of residual heat from planetary accretion, heat produced through radioactive decay, latent heat from core crystallization, and possibly heat from other sources. The major heat-producing isotopes in the Earth are potassium-40, uranium-238, uranium-235, and thorium-232. At the center of the planet, the temperature may be up to 7,000 K and the pressure could reach 360 GPa (3.6 million atm). Because much of the heat is provided by radioactive decay, scientists believe that early in Earth history, before isotopes with short half-lives had been depleted, Earth's heat production would have been much higher. Heat production was twice that of present-day at approximately 3 billion years ago, resulting in larger temperature gradients within the Earth, larger rates of mantle convection and plate tectonics, allowing the production of igneous rocks such as komatiites that are no longer formed.

Heat Sources

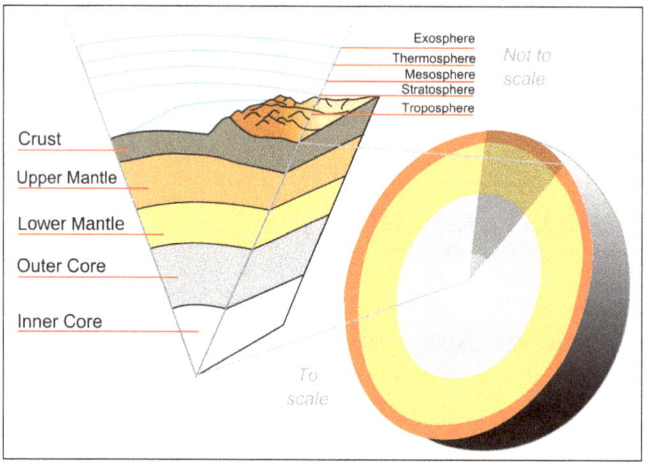

Earth cutaway from core to exosphere.

Temperature within the Earth increases with depth. Highly viscous or partially molten rock at temperatures between 650 to 1,200 °C (1,200 to 2,200 °F) are found at the margins of tectonic plates, increasing the geothermal gradient in the vicinity, but only the outer core is postulated to exist in a molten or fluid state, and the temperature at the Earth's inner core/outer core boundary, around 3,500 kilometres (2,200 mi) deep, is estimated to be 5650 ± 600 Kelvin. The heat content of the Earth is 10^{31} joules.

- Much of the heat is created by decay of naturally radioactive elements. An estimated 45 to 90 percent of the heat escaping from the Earth originates from radioactive decay of elements mainly located in the mantle.

Geothermal drill machine.

- Gravitational potential energy released during the accretion of the Earth.
- Heat released during differentiation, as abundant heavy metals (iron, nickel, copper) descended to the Earth's core.
- Latent heat released as the liquid outer core crystallizes at the inner core boundary.
- Heat may be generated by tidal forces on the Earth as it rotates. The resulting earth tides dissipate energy in Earth's interior as heat.

- There is no reputable science to suggest that any significant heat may be created by the Earth's magnetic field, as suggested by some contemporary folk theories.

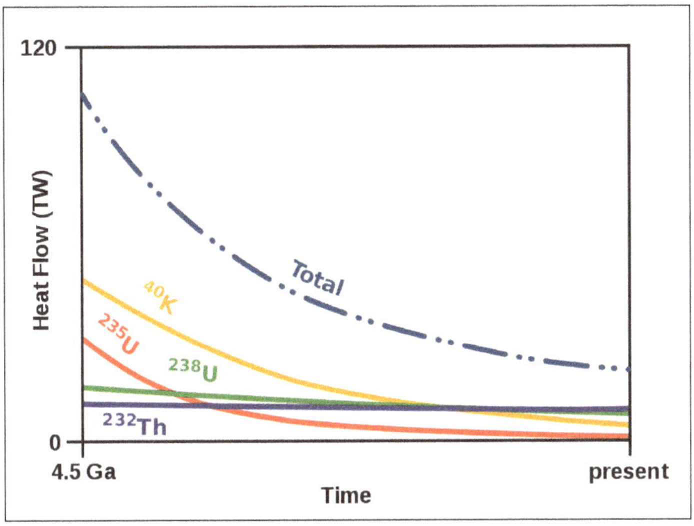

The radiogenic heat from the decay of ^{238}U and ^{232}Th are now the major contributors to the earth's internal heat budget.

In Earth's continental crust, the decay of natural radioactive isotopes makes a significant contribution to geothermal heat production. The continental crust is abundant in lower density minerals but also contains significant concentrations of heavier lithophilic minerals such as uranium. Because of this, it holds the most concentrated global reservoir of radioactive elements found in the Earth. Especially in layers closer to Earth's surface, naturally occurring isotopes are enriched in the granite and basaltic rocks. These high levels of radioactive elements are largely excluded from the Earth's mantle due to their inability to substitute in mantle minerals and consequent enrichment in melts during mantle melting processes. The mantle is mostly made up of high density minerals with higher concentrations of elements that have relatively small atomic radii such as magnesium (Mg), titanium (Ti), and calcium (Ca).

The geothermal gradient is steeper in the lithosphere than in the mantle because the mantle transports heat primarily by convection, leading to a geothermal gradient that is determined by the mantle adiabat, rather than by the conductive heat transfer processes that predominate in the lithosphere, which acts as a thermal boundary layer of the convecting mantle.

Heat Flow

Heat flows constantly from its sources within the Earth to the surface. Total heat loss from the Earth is estimated at 44.2 TW (4.42×10^{13} Watts). Mean heat flow is 65 mW/m^2 over continental crust and 101 mW/m^2 over oceanic crust. This is 0.087 watt/square meter on average (0.03 percent of solar power absorbed by the Earth), but is much more concentrated in areas where the lithosphere is thin, such as along mid-ocean ridges (where new oceanic lithosphere is created) and near mantle plumes. The Earth's crust effectively acts as a thick insulating blanket which must be pierced by fluid conduits (of magma, water or other) in order to release the heat underneath. More of the heat in the Earth is lost through plate tectonics, by mantle upwelling associated with mid-ocean ridges. The final major mode of heat loss is by conduction through the lithosphere, the majority of which occurs in the oceans due to the crust there being much thinner and younger than under the continents.

The heat of the Earth is replenished by radioactive decay at a rate of 30 TW. The global geothermal flow rates are more than twice the rate of human energy consumption from all primary sources.

Direct Application

Heat from Earth's interior can be used as an energy source, known as geothermal energy. The geothermal gradient has been used for space heating and bathing since ancient Roman times, and more recently for generating electricity. As the human population continues to grow, so does energy use and the correlating environmental impacts that are consistent with global primary sources of energy. This has caused a growing interest in finding sources of energy that are renewable and have reduced greenhouse gas emissions. In areas of high geothermal energy density, current technology allows for the generation of electrical power because of the corresponding high temperatures. Generating electrical power from geothermal resources requires no fuel while providing true baseload energy at a reliability rate that constantly exceeds 90%. In order to extract geothermal energy, it is necessary to efficiently transfer heat from a geothermal reservoir to a power plant, where electrical energy is converted from heat by passing steam through a turbine connected to a generator. On a worldwide scale, the heat stored in Earth's interior provides an energy that is still seen as an exotic source. About 10 GW of geothermal electric capacity is installed around the world as of 2007, generating 0.3% of global electricity demand. An additional 28 GW of direct geothermal heating capacity is installed for district heating, space heating, spas, industrial processes, desalination and agricultural applications.

Variations

The geothermal gradient varies with location and is typically measured by determining the bottom open-hole temperature after borehole drilling. Temperature logs obtained immediately after drilling are however affected due to drilling fluid circulation. To obtain accurate bottom hole temperature estimates, it is necessary for the well to reach stable temperature. This is not always achievable for practical reasons.

In stable tectonic areas in the tropics a temperature-depth plot will converge to the annual average surface temperature. However, in areas where deep permafrost developed during the Pleistocene a low temperature anomaly can be observed that persists down to several hundred metres. The Suwałki cold anomaly in Poland has led to the recognition that similar thermal disturbances related to Pleistocene-Holocene climatic changes are recorded in boreholes throughout Poland, as well as in Alaska, northern Canada, and Siberia.

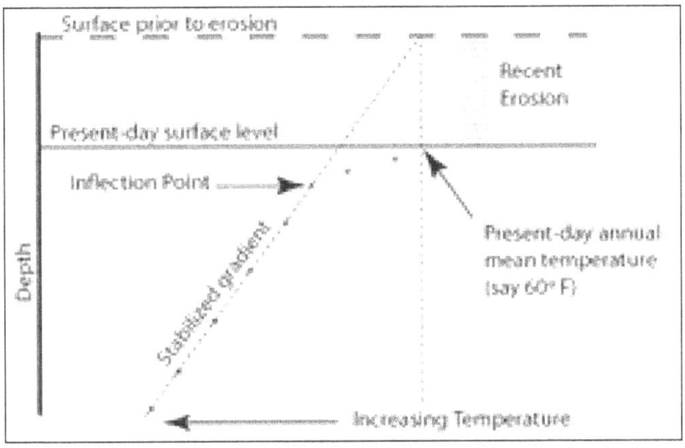

Borehole geothermal gradient in an area of uplift and erosion.

In areas of Holocene uplift and erosion the shallow gradient will be high until it reaches an inflection point where it reaches the stabilized heat-flow regime. If the gradient of the stabilized regime is projected above the inflection point to its intersect with present-day annual average temperature, the height of this intersect above present-day surface level gives a measure of the extent of Holocene uplift and erosion. In areas of Holocene subsidence and deposition the initial gradient will be lower than the average until it reaches an inflection point where it joins the stabilized heat-flow regime.

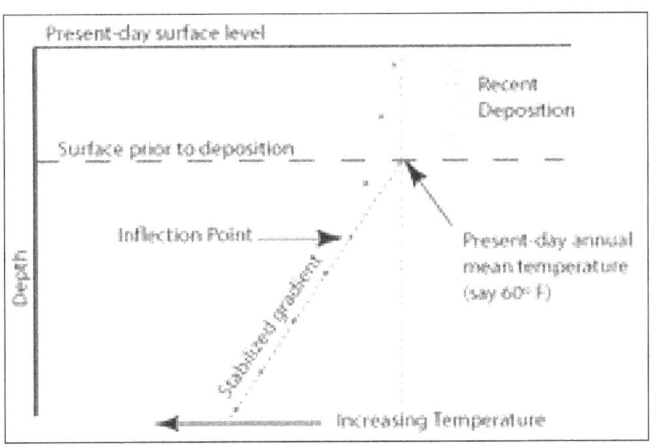

Borehole geothermal gradient in an area of deposition and subsidence.

A variation in surface temperature induced by climate changes and the Milankovitch cycle can penetrate below the Earth's surface and produce an oscillation in the geothermal gradient with periods varying from daily to tens of thousands of years and an amplitude which decreases with depth and having a scale depth of several kilometers. Melt water from the polar ice caps flowing along ocean bottoms tends to maintain a constant geothermal gradient throughout the Earth's surface.

If the rate of temperature increase with depth observed in shallow boreholes were to persist at greater depths, temperatures deep within the Earth would soon reach the point where rocks would melt. We know, however, that the Earth's mantle is solid because of the transmission of S-waves. The temperature gradient dramatically decreases with depth for two reasons. First, the mechanism of thermal transport changes from conduction, as within the rigid tectonic plates, to convection, in the portion of Earth's mantle that convects. Despite its solidity, most of the Earth's mantle behaves over long time-scales as a fluid, and heat is transported by advection, or material transport. Second, radioactive heat production is concentrated within the crust of the Earth, and particularly within the upper part of the crust, as concentrations of uranium, thorium, and potassium are highest there: these three elements are the main producers of radioactive heat within the Earth. Thus, the geothermal gradient within the bulk of Earth's mantle is of the order of 0.5 kelvin per kilometer, and is determined by the adiabatic gradient associated with mantle material (peridotite in the upper mantle).

Mineral Physics

Mineral physics is the science of materials that compose the interior of planets, particularly the Earth. It overlaps with petrophysics, which focuses on whole-rock properties. It provides information that allows interpretation of surface measurements of seismic waves, gravity anomalies, geomagnetic fields and electromagnetic fields in terms of properties in the deep interior of the Earth. This information can be used to provide insights into plate tectonics, mantle convection, the geodynamo and related phenomena.

Laboratory work in mineral physics require high pressure measurements. The most common tool is a diamond anvil cell, which uses diamonds to put a small sample under pressure that can approach the conditions in the Earth's interior.

Creating High Pressures

Shock Compression

Many of the pioneering studies in mineral physics involved explosions or projectiles that subject a sample to a shock. For a brief time interval, the sample is under pressure as the shock wave passes through. Pressures as high as any in the Earth have been achieved by this method. However, the method has some disadvantages. The pressure is very non-uniform and is not adiabatic, so the pressure wave heats the sample up in passing. The conditions of the experiment must be interpreted in terms of a set of pressure-density curves called Hugoniot curves.

Multi-anvil Press

Multi-anvil presses involve an arrangement of anvils to concentrate pressure from a press onto a sample. Typically the apparatus uses an arrangement eight cube-shaped tungsten carbide anvils to compress a ceramic octahedron containing the sample and a ceramic or Re metal furnace. The anvils are typically placed in a large hydraulic press. The method was developed by Kawai and Endo in Japan. Unlike shock compression, the pressure exerted is steady, and the sample can be heated using a furnace. Pressures of about 28 GPa (equivalent to depths of 840 km), and temperatures above 2300 °C, can be attained using WC anvils and a lanthanum chromite furnace. The apparatus is very bulky and cannot achieve pressures like those in the diamond anvil cell, but it can handle much larger samples that can be quenched and examined after the experiment. Recently, sintered diamond anvils have been developed for this type of press that can reach pressures of 90 GPa (2700 km depth).

Diamond Anvil Cell

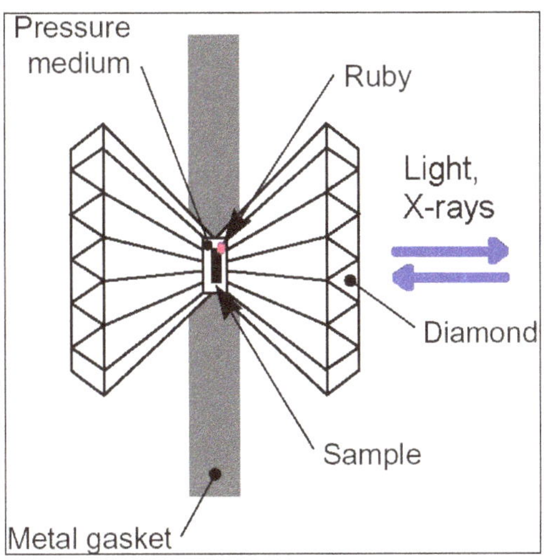

Schematics of the core of a diamond anvil cell. The diamond size is a few millimeters at most.

The diamond anvil cell is a small table-top device for concentrating pressure. It can compress a small (sub-millimeter sized) piece of material to extreme pressures, which can exceed 3,000,000 atmospheres (300 gigapascals). This is beyond the pressures at the center of the Earth. The concentration of pressure at the tip of the diamonds is possible because of their hardness, while their transparency and high thermal conductivity allow a variety of probes can be used to examine the state of the sample. The sample can be heated to thousands of degrees.

Creating High Temperatures

Achieving temperatures found within the interior of the earth is just as important to the study of mineral physics as creating high pressures. Several methods are used to reach these temperatures and measure them. Resistive heating is the most common and simplest to measure. The application of a voltage to a wire heats the wire and surrounding area. A large variety of heater designs are available including those that heat the entire diamond anvil cell (DAC) body and those that fit inside the body to heat the sample chamber. Temperatures below 700°C can be reached in air due to the oxidation of diamond above this temperature. With an argon atmosphere, higher temperatures up to 1700°C can be reached without damaging the diamonds. Resistive heaters have not achieved temperatures above 1000°C.

Laser heating is done in a diamond-anvil cell with Nd:YAG or CO2 lasers to achieve temperatures above 6000k. Spectroscopy is used to measure black-body radiation from the sample to determine the temperature. Laser heating is continuing to extend the temperature range that can be reached in diamond-anvil cell but suffers two significant drawbacks. First, temperatures below 1200°C are difficult to measure using this method. Second, large temperature gradients exist in the sample because only the portion of sample hit by the laser is heated.

Properties of Materials

Equations of State

To deduce the properties of minerals in the deep Earth, it is necessary to know how their density varies with pressure and temperature. Such a relation is called an equation of state (EOS). A simple example of an EOS that is predicted by the Debye model for harmonic lattice vibrations is the Mie-Grünheisen equation of state:

$$\left(\frac{dP}{dT}\right)=\frac{\gamma_D}{V}C_V,$$

where C_V is the heat capacity and γ_D is the Debye gamma. The latter is one of many Grünheisen parameters that play an important role in high-pressure physics. A more realistic EOS is the Birch–Murnaghan equation of state.

Interpreting Seismic Velocities

Inversion of seismic data give profiles of seismic velocity as a function of depth. These must still be interpreted in terms of the properties of the minerals. A very useful heuristic was discovered by Francis Birch: plotting data for a large number of rocks, he found a linear relation of the compressional wave velocity v_p of rocks and minerals of a constant average atomic weight $\bar{M}$ with density ρ:

$$v_p = a\bar{M} + b\rho.$$

This relationship became known as Birch's law. This makes it possible to extrapolate known velocities for minerals at the surface to predict velocities deeper in the Earth.

Other Physical Properties

- Viscosity.
- Creep (deformation).
- Melting.
- Electrical conduction and other transport properties.

Methods of Crystal Interrogation

There are a number of experimental procedures designed to extract information from both single and powdered crystals. Some techniques can be used in a diamond anvil cell (DAC) or a multi anvil press (MAP). Some techniques are summarized in the following table.

Technique	Anvil Type	Sample Type	Information Extracted	Limitations
X-ray Diffraction(XRD)	DAC or MAP	Powder or Single Crystal	Cell parameters	
Electron Microscopoy	Neither	Powder or Single Crystal	Symmetry Group	Surface Measurements Only
Neutron Diffraction	Neither	Powder	Cell parameters	Large Sample needed
Infrared spectroscopy	DAC	Powder, Single Crystal or Solution	Chemical Composition	Not all materials are IR active
Raman Spectroscopy	DAC	Powder, Single Crystal or Solution	Chemical Composition	Not all materials are Raman active
Brillouin Scattering	DAC	Single Crystal	Elastic Moduli	Need optically thin sample
Ultrasonic Interferometry	DAC or MAP	Single Crystal	Elastic Moduli	

First Principles Calculations

Using quantum mechanical numerical techniques, it is possible to achieve very accurate predictions of crystal's properties including structure, thermodynamic stability, elastic properties and transport properties. The limit of such calculations tends to be computing power, as computation run times of weeks or even months are not uncommon.

Dynamic Topography

Dynamic topography is that topography due to flow within the mantle, an idea that was first proposed by Pekeris. It is "dynamic" because the mass anomalies driving the density are moving, and this contrasts with isostatic topography, in which the mass anomalies are in a state of quasi-equilibrium. Due to plate tectonics, the topography of Earth consists of continents and oceans with an average elevation difference of ~4.5 km. This first-order difference in elevation is of isostatic origin and results from the different average densities and thicknesses of continental and oceanic crust and lithosphere. The long-known poor correlation between the ocean-continent function and the long-wavelength geoid supports this view. If the topography of Earth were not affected by any dynamic processes, flat continents would stand ~4.5 km above flat abyssal plains. This is clearly not the case, and the topography of Earth results from an array of dynamic processes operating at different scales in time and space. In the continents, tectonic convergence results in mountain belts such as the Himalayas and the Andes, and tectonic divergence results in rifts, such as the East African Rift or the Salton Sea in California. The process leading to these topographic features is clearly dynamic, but even in active and geologically recent tectonic zones, the topography is largely compensated isostatically: The continental crust is thicker than average in mountain belts, and thinner than average in rifts.

In the oceans, the bathymetry is primarily determined by cooling and thickening of oceanic lithosphere as it moves away from spreading centers. While it is ultimately dynamic in origin, the bathymetry of the young ocean floor (<80 Ma) is well explained by combining the principles of isostasy and of the cooling of a semi-infinite half space. Ocean floor older than ca. 80 Ma appears to flatten compared to this simple and elegant physical model, although whether "normal" old seafloor flattens is subject to some debate. The reason for the apparent shallowing of old ocean floor is the subject of ongoing debate, and proposed explanations include thermal rejuvenation by hotspots, small-scale convection, dynamic support by mantle flow, and departure from the topography predicted by thermal boundary layer theory due to radioactive heat production within the mantle. In the absence of consensus regarding the physical process explaining this flattening, the age-depth relationship of the ocean floor can still be accommodated in empirical "plate models". As in the continents, the topography of the ocean floor is affected by variations in crustal thickness, notably in areas affected by magmatism such as oceanic plateaus. Similarly, variations in lithospheric temperature and magmatic processes also affect continental topography; however, there is no simple tectonic model to describe the topography of continents due to their long and complex tectonic history.

Given the number of dynamic processes that affect topography, it is not surprising that the term "dynamic topography," which generally describes the effect of mantle convection on surface topography, does not have a single, agreed-upon definition. The bathymetry of the ocean floor is compatible with that of the thermal boundary layer of the convecting mantle, and, in this respect, the subsidence of oceanic lithosphere with age should be included in the definition of dynamic topography. However, the dynamics of the continental lithosphere are more complex and cannot be explained by boundary layer theory. For this reason, defining dynamic topography globally as the difference between observed topography and isostatic crustal topography results in continents dynamically depressed by up to 3 km, which is inconsistent with observations. This discrepancy is in part due to the non-negligible isostatic contribution of the continental lithosphere to topography. Given the relatively poor global constraints on the complex structure and density of the continental lithosphere, dynamic topography is usually defined as that topography originating from sources beneath the upper thermal boundary layer of mantle convection.

Residual Topography

Constraining dynamic topography requires removing the isostatic contribution of the sediments, ice, crust, and lithosphere from the observed topography. The remaining topography is called residual topography. In the oceans, a choice must be made between correcting either for the physical half-space cooling model or for a plate model accounting for the apparent flattening of the ocean floor with increasing age. The former approach implies that mantle support may contribute to the observed flattening of old ocean floor, whereas the latter requires some other physical process to operate. In the continents, a common approach is to remove the isostatic topography due to the continental crust. However, correcting for the isostatic contribution of the continental crust alone can result in continents being either systematically lowered or elevated by 1 to 2 km, which is inconsistent with observation. Zoback and Mooney showed that there is no simple relationship between continental crustal thickness and surface elevation, and that isostatic models based on crustal thickness alone systematically overpredict continental elevations, mainly because Archean continental lithosphere is depleted and of intrinsically lower density. Several authors have proposed residual topography maps that remove the isostatic contribution of both the continental crust and continental lithosphere. Such calculations require assumptions relative to the thermo-chemical state of the continental lithosphere, which result in relatively large uncertainties and underline the need for a better knowledge of the thermal and chemical structure of the lithosphere in order to constrain mantle flow models.

The long-wavelength residual topography has been calculated by removing the mean continental elevation from continents, because the thermal and chemical structures of the continental lithosphere are not well known. In the oceans, the plate cooling model of Crosby and McKenzie was removed from the observed bathymetry after accounting for isostatic loading by sediments using the method of Sykes and the global sediment thickness map of Divins. We consider that continents extend to the edge of the continental shelf, which roughly corresponds to the −200 m elevation contour, and we calculate the equal-area arithmetic mean continental elevation as 529 m. The areas between the edge of the continental shelf and ocean floor per se as defined by were arbitrarily set to a uniform elevation of −200 m. The resulting residual topography was low-pass filtered using a filter tapered for wavelength between 2000 and 5000 km, in order to simplify comparison with published estimates of residual topography expanded in spherical harmonics to degree 12. The residual topography contains an isostatic component evident in high values in the Himalayas and Andes (with present crustal thicknesses greater than 50 km. In the oceans, the residual topography highs are closely associated with hotspot tracks and Phanerozoic large igneous provinces Therefore, part of the oceanic residual topography is due to the thicker crust of oceanic plateaus and at least partially isostatically compensated. Ideally, a correction for the thickness of the oceanic crust should be carried out to estimate the topography associated with mantle support. Such studies have been carried out regionally but not yet globally. Finally, the residual topography high over the Canada Basin in the Arctic Ocean can be attributed to the absence of sediment thickness data.

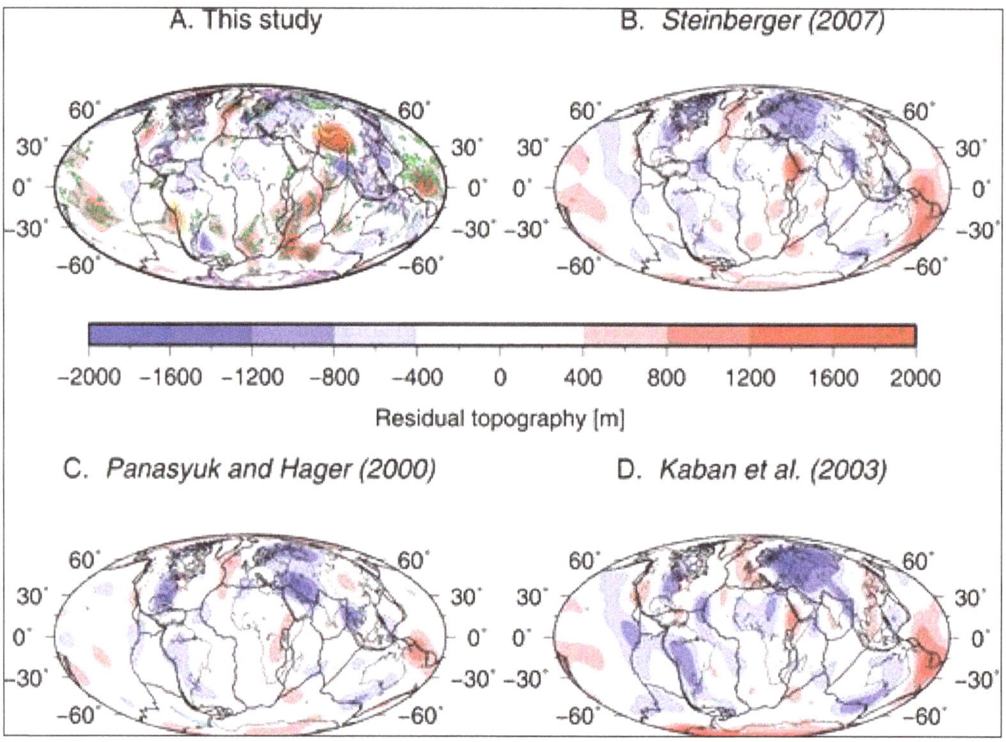

Graphical comparison between four residual topography fields calculated independently. All four models account for the flattening of old ocean floor ("plate model") with water-loaded oceans. Continents are air-loaded, and the mean residual topography is set to zero for comparison with dynamic topography models. In the continents, A was calculated by removing the mean continental elevation (529 m), B was obtained by removing the isostatic contribution of the crust only, updated from Steinberger using the age grids of Müller et al., and C and D were obtained by removing the isostatic contribution of both the continental crust and the continental lithosphere. The thin and thick black lines are the coastlines and plate boundaries, respectively. In A, the white contours are the continent-ocean boundary from Müller et al., the magenta contours are the edge of the continents, the green contours outline Phanerozoic large igneous provinces, and the yellow contours indicate continental crust thicker than 50 km. The red star is the location of well COST-B2 offshore New Jersey.

Despite being based on different isostatic models of the lithosphere, the four residual topography fields agree overall. For half-wavelengths greater than ~3300 km, the peak-to-peak amplitude is ±2 km, and the root-mean-square amplitude is ~0.5 km, due to the absence of correcting for isostasy in elevated areas. The root-mean-square amplitude is smallest for the model of Panasyuk and Hager and largest for the model of Kaban et al. In terms of pattern, in the continents, features common to all four models include the residual topography lows within central Europe, eastern North America, and western South America, and the residual topography highs within southern and eastern Africa and western North America. In the oceans, the most persistent features between models are the Australian-Antarctic Discordance (AAD) that depresses the southeast Indian mid-ocean ridge, the Argentine Basin in the southern South Atlantic, the residual topography low under the western part of the central Atlantic Ocean, the large-scale, low-amplitude residual topography low under the northeast Pacific Ocean, and the highs under the central and western Pacific Ocean (Darwin Rise), the areas offshore southern Africa, and around Iceland and the northern Atlantic mid-ocean ridge.

Time-dependent Constraints on Dynamic Topography

Dynamic topography can occur over all wavelengths from 10 km to greater than 1000 km, but since crustal and lithospheric density and thickness variations and plate flexure exert a strong control on topography for wavelengths less than 100 km, most attempts to deconvolve it from the geological record have focused on transient long-wavelength changes that cannot be well explained without invoking dynamic topography.

Geological Expressions of Long-wavelength Dynamic Topography

Long-wavelength dynamic topography is recorded by lithospheric plates moving over the constantly changing mantle flow and is thus ephemeral over millions of years to tens of millions of years. As a result, subsidence events are generally followed by uplift and vice versa. Indeed, the fact that dynamic topography is transient over tens of millions of years means that it must be recorded over long time periods, over which the preservation potential of the rock record is poor. The very nature of the long-wavelength, low-amplitude dynamic topography signal makes it difficult to isolate in the geological record, which is dominated by shorter-wavelength, larger-amplitude signals such as lithospheric extension or shortening. One further restriction on the preservation of dynamic topography comes from the geological record itself, which is biased toward areas that have been subsiding for long periods of time and have a good preservation potential, as opposed to uplifting areas that are subject to erosion and have a poor preservation potential. However, strata deposited in a dynamic topography depression also have low long-term preservation potential because dynamic topography is transient. Thus, ancient subduction-related dynamic topography is most likely to be represented by unconformities. As a consequence, the stratigraphic record is the most accessible archive of the dynamic topography signal, and continental interiors that have not undergone deformation for hundreds of millions of years are the most promising places to search for the dynamic topography signal.

References

- Stern, R.J. (14 August 2007). Modern-style plate tectonics began in neoproterozoic time: an alternative interpretation of Earth's tectonic history. Geological Society of America Special Paper. 440. pp. 265–280. doi:10.1130/2008.2440(13). ISBN 978-0-8137-2440-9
- Geologia-de-Engenharia-Topografia, Biblio, Deformation-of-Rock: 3.uma.pt, Retrieved 11 February, 2020
- Korenaga, J. (10 December 2010). "On the Likelihood of Plate Tectonics on Super-Earths: Does Size Matter?". The Astrophysical Journal Letters. 725 (1): 43–46. Bibcode:2010ApJ...725L..43K. doi:10.1088/2041-8205/725/1/L43
- A-review-of-observations-and-models-of-dynamic, lithosphere-145653: pubs.geoscienceworld.org, Retrieved 03 May, 2020
- Price, G. David (October 2007). "2.01 Overview – Mineral physics: past, present, and future" (PDF). In Price, G. David (ed.). Mineral Physics. Elsevier. pp. 1–6. ISBN 9780444535764. Retrieved 27 September 2017

Geomagnetism and Paleomagnetism 4

- Geomagnetism
- Magnetic Declination
- Magnetic Inclination
- Dipole Model of the Earth's Magnetic Field
- The Dynamo Effect
- Geomagnetic Reversal
- Magnetic Anomaly
- Paleomagnetism
- Principles of Remanent Magnetization
- Paleomagnetic Poles

The dynamics of the Earth's magnetic field are studied under geomagnetism. Paleomagnetism is associated with the magnetism in rocks induced by the Earth's magnetic field at the time of formation. The topics elaborated in this chapter will help in gaining a better perspective of the concepts related to geomagnetism and paleomagnetism.

Geomagnetism

Geomagnetism is the study of the dynamics of the Earth's magnetic field, which is produced in the inner core. The Earth's magnetic field is predominantly a geo-axial dipole, with north and south magnetic poles located near the geographic poles that undergo periodic reversals and excursions.

Evidence

Geomagnetism and geomagnetic reversals are supported by several threads of evidence.

Magnetic Seafloor Stripes

During the 1950s-70s, scientists learned that the ocean seafloor is patterned by stripes of magnetism, with alternating sections magnetized north, south, north, etc.

This occurs because magma contains iron grains (specifically magnetite) that orient themselves towards the magnetic north pole, much like a compass. When lava solidifies, the iron grains are no longer able to move freely, and are permanently aligned towards wherever the magnetic north pole was. Thus, when magma from oceanic trenches hardens into seafloor, it will maintain its magnetic alignment. Because stripes of the same number of magnetic reversals away from oceanic trenches are uniform in age, it is clear that this is a global event, which can only be explained by something like a geomagnetic reversal.

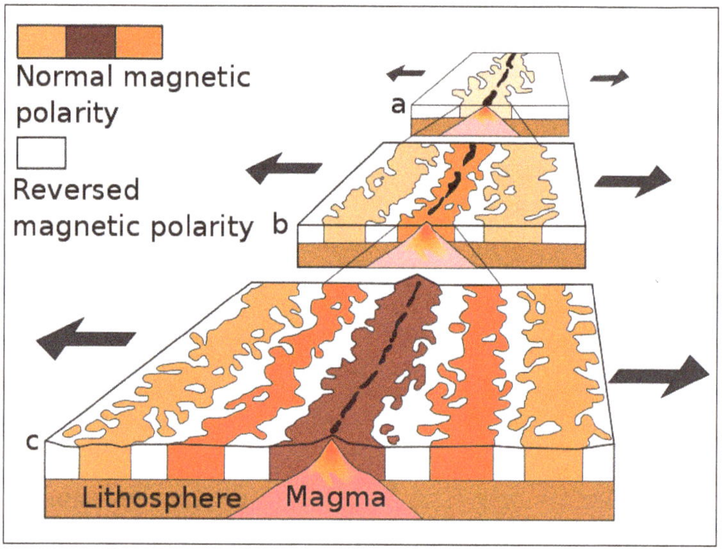

The seafloor spreading process.

The theory of seafloor spreading is supported by the existence of symmetrical (in age, polarity, and distance from mid-ocean trenches) magnetic stripes in the rock of the ocean floor. No other theory can explain why these stripes exist at all *or* their symmetry, making this strong evidence for seafloor spreading. This, in turn, supports the existence of geomagnetic reversals.

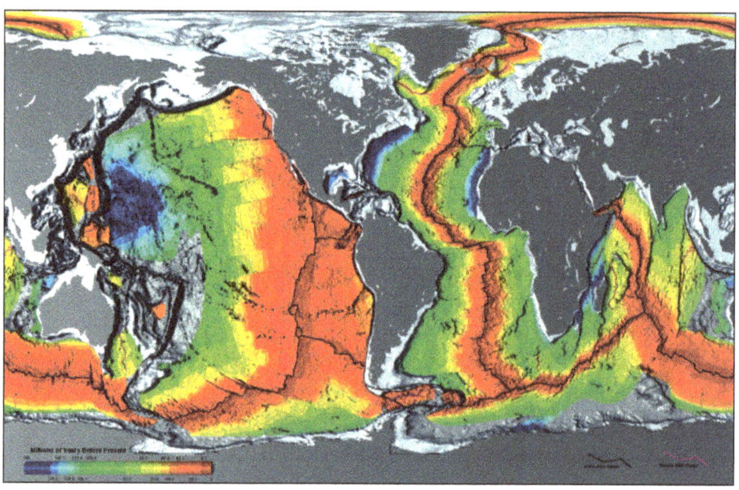

Earth's seafloor crust ages.

Geodynamo Theory

If the Earth's magnetic field was freely decaying (nothing added to its strength, so it slowly died) then it would decay in about 20,000 years. Since we know that the magnetic field has existed for at least 3,000,000,000 years, a theory is needed to explain how the Earth's field regenerates that also can explain the existence of geomagnetic reversals, local fluctuations in the geomagnetic field, and various other magnetic phenomena.

3D simulation of the Earth's magnetism.

This led to the rise of the geodynamo theory. The geodynamo theory has two central ideas:

- The Earth's core is filled with moving, liquid metal that is constantly flowing. Heavier and cooler metals settle onto the Earth's core, while lighter and hotter metals rise, making a gigantic convection current, which, due to its metallic contents, gives rise to a magnetic field.
- The Earth spins on its axis, which puts Coriolis forces onto the core, which forces it to spin and to shear through the previous magnetic field, which generates a new one, which keeps the field as a whole alive.

One model of the geodynamo theory, the Glatzmaier-Roberts model, was simulated for more than 300,000 (simulated) years. The simulated field matched the observed field well. Notably:

- The simulated field's strength matched the observed field's strength.
- The simulated field's westward pole drift matched the observed field's westward pole drift in direction.
- The simulated field had polar reversals in ~100,000 year intervals that matched the observed field's reversal timeframe.

Impacts on Life

Because the strength of the geomagnetic field is currently declining, it has become the center of some woo. It is believed that a reversal or excursion (basically a reversal, with associated decrease of strength, minus the polar flip) may be imminent in the next couple of thousand of years or so.

Reversals *do* have an impact. For example, during the most recent reversal, the Earth's magnetic field reduced to 5% of its original strength. A small amount of Earth's atmosphere, usually protected by the Earth's geomagnetic field, can be lost to space due to radioactive bombardment from the sun. On the whole, however, reversals appear to be relatively harmless to living beings. Considering that around 170 to 171 reversals are documented in the last 76 million years alone, reversals don't seem to pose much of an extinction-level threat.

Magnetic Apocalypse

Some cranks, including Robert W. Felix, believe that geomagnetic reversals cause ice ages and extinctions that lead to "evolutionary leaps". While both are possible, both are rather unlikely. Geomagnetic reversals don't seem to have impacted life on Earth much in the past 76 million years.

Pole Shift

A geomagnetic reversal is *not* a pole shift. In a geomagnetic reversal, the magnetic poles are flipped. In a pole shift, the planet itself is flipped the Arctic located where the Antarctic is, and vice versa.

Impact Theory

In "Impact theory", it has been proposed that very large meteor impacts may cause or may contribute to geomagnetic reversals, because:

- Very large meteor impacts throw up enough dust to cover most of the Earth's surface from the sun.
- If enough dust is thrown up, the Earth's surface water may freeze.
- Frozen water flows differently than liquid water, which affects the spinning of the earth's crust and mantle.
- The spinning of the crust and mantle is what generates the geodynamo, and changing it will cause new patterns to form.

Creationist Interpretations

Geomagnetic reversals, given their known larger-than-10,000-years timeframe and their support for much-slower-than-10,000-years seafloor spreading and plate tectonics, present a problem for young earth creationism.

Decay Only

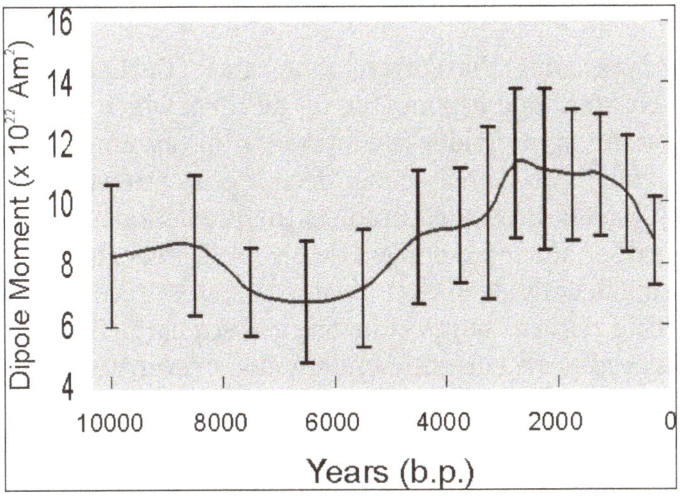

Estimated dipole moment (basically, the strength of the magnetic field between the two poles) of the Earth from 10000 years before present.

Some young earth creationists, such as Walt Brown, Thomas Barnes, Kent Hovind, the ICR, and CMI's 101 evidences for a young age of the Earth and the universe (numbers 49, 69, 70, 71, and 81) argue that the geomagnetic field has only (or with slight variation in speed) decayed for the past 6-10,000 years; geomagnetic reversals would contradict this worldview. Depending on the level of denialism, many creationists either deny that the data exists or consider them to be past fluctuations in the overall high strength of the geomagnetic field.

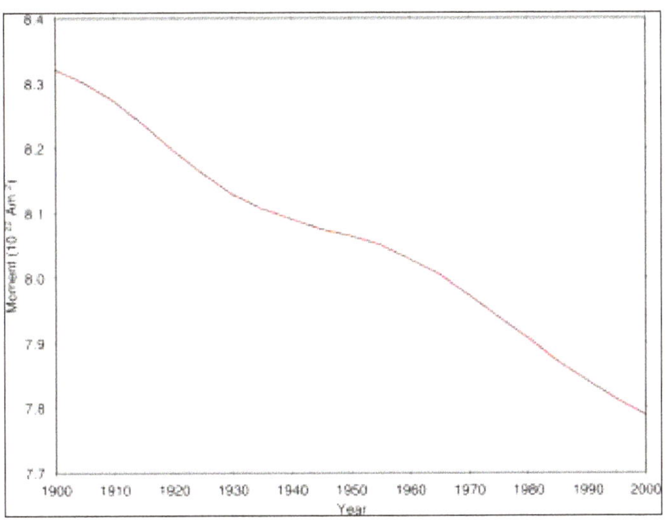

Even excluding the above graph, this looks less like exponential decay and lot more like *linear* decay.

Why? Because, based on the decline of the magnetic field in the past century or two, they believe that the magnetic field has been in constant, exponential decline, forever. Extrapolating exponentially backwards, YECs conclude that the Earth cannot be more than 20,000 years old, because that would imply that the Earth's magnetic field would be impossibly large. This assertion has numerous problems:

- First, this idea assumes that current decay rates of the Earth's magnetic field have remained constant and exponential for all time, which can be seen in the lower image on the right. Evidence contradicts this assumption. The geomagnetic field frequently both increases and decreases in strength, with no clear trend either way, which directly contradicts the predictions of this model. Further, as shown earlier, during geomagnetic reversals and excursions, the strength of the field significantly decreases; during the most recent reversal, the Earth's magnetic field reduced to 5% of its original strength. The Earth's geomagnetic field is believed to be currently undergoing a reversal or an excursion, which would account for the decrease of the strength of the field. This means that the decay rate cannot be assumed to be, in fact, constant, and thus the proposed timeframe of this model cannot be accepted.
- Second, this model relies on an obsolete model of the Earth's geomagnetism. In order for it to make sense, it must assume that the source of geomagnetism is a free electrical current in the Earth's molten center. (If we accept this, then 20,000 years is the time in which the Earth's magnetic field would have decayed away.)

Geomagnetic Field

Geomagnetic field is the magnetic field that extends from the Earth's interior out into space, where it interacts with the solar wind, a stream of charged particles emanating from the Sun. The magnetic field is generated by electric currents due to the motion of convection currents of molten iron in the Earth's outer core: these convection currents are caused by heat escaping from the core, a natural process called a geodynamo. The magnitude of the Earth's magnetic field at its surface ranges from 25 to 65 microteslas (0.25 to 0.65 gauss). As an approximation, it is represented by a field of a magnetic dipole currently tilted at an angle of about 11 degrees with respect to Earth's rotational axis, as if there were an enormous bar magnet placed at that angle through the center of the Earth. The North geomagnetic pole, which was in 2015 located on Ellesmere Island, Nunavut, Canada, in the northern hemisphere, is actually the south pole of the Earth's magnetic field, and conversely.

While the North and South magnetic poles are usually located near the geographic poles, they slowly and continuously move over geological time scales, but sufficiently slowly for ordinary compasses to remain useful for navigation. However, at irregular intervals averaging several hundred thousand years, the Earth's field reverses and the North and South Magnetic Poles respectively, abruptly switch places. These reversals of the geomagnetic poles leave a record in rocks that are of value to paleomagnetists in calculating geomagnetic fields in the past. Such information in turn is helpful in studying the motions of continents and ocean floors in the process of plate tectonics.

The magnetosphere is the region above the ionosphere that is defined by the extent of the Earth's magnetic field in space. It extends several tens of thousands of kilometers into space, protecting the Earth from the charged particles of the solar wind and cosmic rays that would otherwise strip away the upper atmosphere, including the ozone layer that protects the Earth from harmful ultraviolet radiation.

The Earth's magnetic field serves to deflect most of the solar wind, whose charged particles would otherwise strip away the ozone layer that protects the Earth from harmful ultraviolet radiation. One stripping mechanism is for gas to be caught in bubbles of magnetic field, which are ripped off by solar winds. Calculations of the loss of carbon dioxide from the atmosphere of Mars, resulting from scavenging of ions by the solar wind, indicate that the dissipation of the magnetic field of Mars caused a near total loss of its atmosphere.

The study of the past magnetic field of the Earth is known as paleomagnetism. The polarity of the Earth's magnetic field is recorded in igneous rocks, and reversals of the field are thus detectable as "stripes" centered on mid-ocean ridges where the sea floor is spreading, while the stability of the geomagnetic poles between reversals has allowed paleomagnetists to track the past motion of continents. Reversals also provide the basis for magnetostratigraphy, a way of dating rocks and sediments. The field also magnetizes the crust, and magnetic anomalies can be used to search for deposits of metal ores.

Humans have used compasses for direction finding since the 11th century A.D. and for navigation since the 12th century. Although the magnetic declination does shift with time, this wandering is slow enough that a simple compass can remain useful for navigation. Using magnetoreception various other organisms, ranging from some types of bacteria to pigeons, use the Earth's magnetic field for orientation and navigation.

Characteristics

At any location, the Earth's magnetic field can be represented by a three-dimensional vector. A typical procedure for measuring its direction is to use a compass to determine the direction of magnetic North. Its angle relative to true North is the *declination* (D) or *variation*. Facing magnetic North, the angle the field makes with the horizontal is the *inclination* (I) or *magnetic dip*. The *intensity* (F) of the field is proportional to the force it exerts on a magnet. Another common representation is in X (North), Y (East) and Z (Down) coordinates.

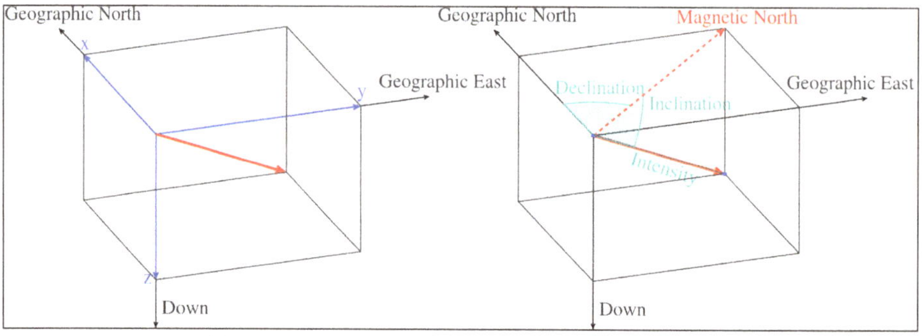

Common coordinate systems used for representing the Earth's magnetic field.

Intensity

The intensity of the field is often measured in gauss (G), but is generally reported in nanoteslas (nT), with 1 G = 100,000 nT. A nanotesla is also referred to as a gamma (γ). The tesla is the SI unit of the magnetic field, The Earth's field ranges between approximately 25,000 and 65,000 nT (0.25–0.65 G). By comparison, a strong refrigerator magnet has a field of about 10,000,000 nanoteslas (100 G).

A map of intensity contours is called an *isodynamic chart*. The intensity tends to decrease from the poles to the equator. A minimum intensity occurs in the South Atlantic Anomaly over South America while there are maxima over northern Canada, Siberia, and the coast of Antarctica south of Australia.

Inclination

The inclination is given by an angle that can assume values between -90° (up) to 90° (down). In the northern hemisphere, the field points downwards. It is straight down at the North Magnetic Pole and rotates upwards as the latitude decreases until it is horizontal (0°) at the magnetic equator. It continues to rotate upwards until it is straight up at the South Magnetic Pole. Inclination can be measured with a dip circle.

Declination

Declination is positive for an eastward deviation of the field relative to true north. It can be estimated by comparing the magnetic north/south heading on a compass with the direction of a celestial pole. Maps typically include information on the declination as an angle or a small diagram showing the relationship between magnetic north and true north. Information on declination for a region can be represented by a chart with isogonic lines (contour lines with each line representing a fixed declination).

Dipolar Approximation

Near the surface of the Earth, its magnetic field can be closely approximated by the field of a magnetic dipole positioned at the center of the Earth and tilted at an angle of about

11° with respect to the rotational axis of the Earth. The dipole is roughly equivalent to a powerful bar magnet, with its south pole pointing towards the geomagnetic North Pole. This may seem surprising, but the north pole of a magnet is so defined because, if allowed to rotate freely, it points roughly northward (in the geographic sense). Since the north pole of a magnet attracts the south poles of other magnets and repels the north poles, it must be attracted to the south pole of Earth's magnet. The dipolar field accounts for 80–90% of the field in most locations.

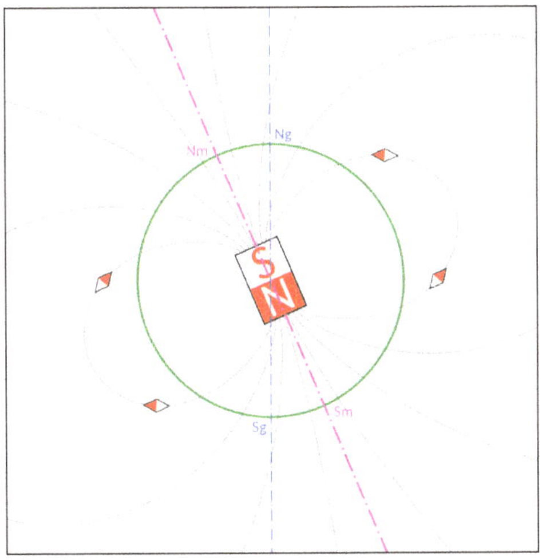

The variation between magnetic north (N_m) and "true" north (N_g).

Magnetic Poles

Historically, the north and south poles of a magnet were first defined by the Earth's magnetic field, not vice versa, since one of the first uses for a magnet was as a compass needle. A magnet's North pole is defined as the pole that is attracted by the Earth's North Magnetic Pole when the magnet is suspended so it can turn freely. Since opposite poles attract, the North Magnetic Pole of the Earth is really the south pole of its magnetic field (the place where the field is directed downward into the Earth).

The positions of the magnetic poles can be defined in at least two ways: locally or globally. The local definition is the point where the magnetic field is vertical. This can be determined by measuring the inclination. The inclination of the Earth's field is 90° (downwards) at the North Magnetic Pole and -90° (upwards) at the South Magnetic Pole. The two poles wander independently of each other and are not directly opposite each other on the globe. Movements of up to 40 kilometres (25 mi) per year have been observed for the North Magnetic Pole. Over the last 180 years, the North Magnetic Pole has been migrating northwestward, from Cape Adelaide in the Boothia Peninsula in 1831 to 600 kilometres (370 mi) from Resolute Bay in 2001. The *magnetic equator* is the line where the inclination is zero (the magnetic field is horizontal).

The global definition of the Earth's field is based on a mathematical model. If a line is drawn through the center of the Earth, parallel to the moment of the best-fitting magnetic dipole, the two positions where it intersects the Earth's surface are called the North and South geomagnetic poles. If the Earth's magnetic field were perfectly dipolar, the geomagnetic poles and magnetic dip poles would coincide and compasses would point towards them. However, the Earth's field has a significant non-dipolar contribution, so the poles do not coincide and compasses do not generally point at either.

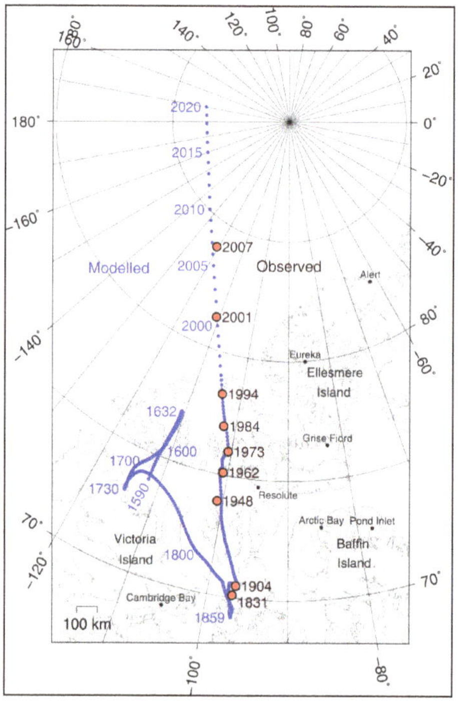

The movement of Earth's North Magnetic Pole across the Canadian arctic.

Magnetosphere

Earth's magnetic field, predominantly dipolar at its surface, is distorted further out by the solar wind. This is a stream of charged particles leaving the Sun's corona and accelerating to a speed of 200 to 1000 kilometres per second. They carry with them a magnetic field, the interplanetary magnetic field (IMF).

The solar wind exerts a pressure, and if it could reach Earth's atmosphere it would erode it. However, it is kept away by the pressure of the Earth's magnetic field. The magnetopause, the area where the pressures balance, is the boundary of the magnetosphere. Despite its name, the magnetosphere is asymmetric, with the sunward side being about 10 Earth radii out but the other side stretching out in a magnetotail that extends beyond 200 Earth radii. Sunward of the magnetopause is the bow shock, the area where the solar wind slows abruptly.

Inside the magnetosphere is the plasmasphere, a donut-shaped region containing low-energy charged particles, or plasma. This region begins at a height of 60 km, extends up to 3 or 4 Earth radii, and includes the ionosphere. This region rotates with the Earth. There are also two concentric tire-shaped regions, called the Van Allen radiation belts, with high-energy ions (energies from 0.1 to 10 million electron volts (MeV)). The inner belt is 1–2 Earth radii out while the outer belt is at 4–7 Earth radii. The plasmasphere and Van Allen belts have partial overlap, with the extent of overlap varying greatly with solar activity.

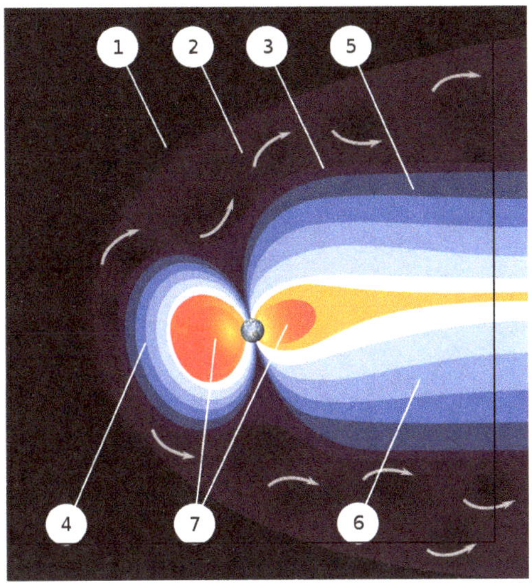

The structure of a magnetosphere. 1) Bow shock. 2) Magnetosheath. 3) Magnetopause. 4) Magnetosphere. 5) Northern tail lobe. 6) Southern tail lobe. 7) Plasmasphere.

As well as deflecting the solar wind, the Earth's magnetic field deflects cosmic rays, high-energy charged particles that are mostly from outside the Solar System. Many cosmic rays are kept out of the Solar System by the Sun's magnetosphere, or heliosphere. By contrast, astronauts on the Moon risk exposure to radiation. Anyone who had been on the Moon's surface during a particularly violent solar eruption in 2005 would have received a lethal dose.

Some of the charged particles do get into the magnetosphere. These spiral around field lines, bouncing back and forth between the poles several times per second. In addition, positive ions slowly drift westward and negative ions drift eastward, giving rise to a ring current. This current reduces the magnetic field at the Earth's surface. Particles that penetrate the ionosphere and collide with the atoms there give rise to the lights of the aurorae and also emit X-rays.

The varying conditions in the magnetosphere, known as space weather, are largely driven by solar activity. If the solar wind is weak, the magnetosphere expands; while if it is strong, it compresses the magnetosphere and more of it gets in. Periods of particularly intense activity, called geomagnetic storms, can occur when a coronal mass ejection erupts above the Sun and sends a shock wave through the Solar System. Such a wave can take just two days to reach the Earth. Geomagnetic storms can cause a lot of disruption; the "Halloween" storm of 2003 damaged more than a third of NASA's satellites. The largest documented storm occurred in 1859. It induced currents strong enough to short out telegraph lines, and aurorae were reported as far south as Hawaii.

Time Dependence

Short-term Variations

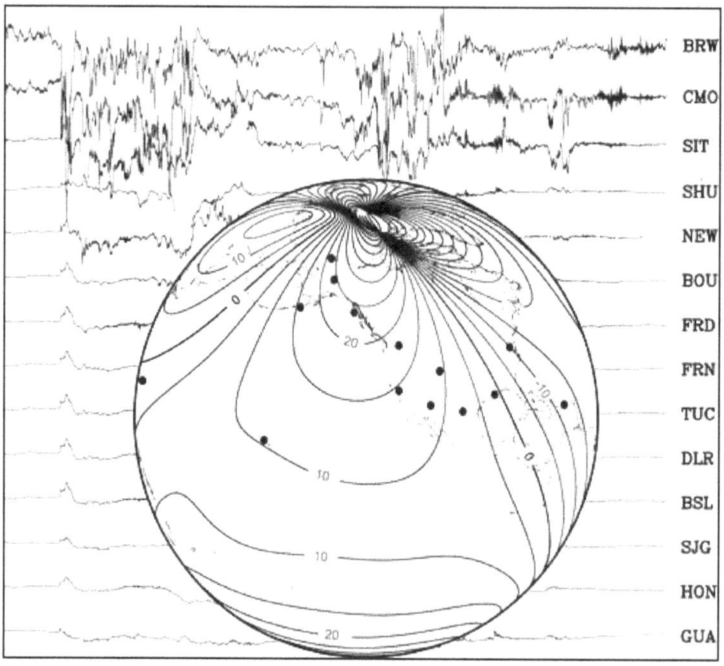

Background: A set of traces from magnetic observatories showing a magnetic storm. Globe: map showing locations of observatories and contour lines giving horizontal magnetic intensity in μ T.

The geomagnetic field changes on time scales from milliseconds to millions of years. Shorter time scales mostly arise from currents in the ionosphere (ionospheric dynamo region) and magnetosphere, and some changes can be traced to geomagnetic storms or daily variations in currents. Changes over time scales of a year or more mostly reflect changes in the Earth's interior, particularly the iron-rich core.

Frequently, the Earth's magnetosphere is hit by solar flares causing geomagnetic storms, provoking displays of aurorae. The short-term instability of the magnetic field is measured with the K-index.

Data from THEMIS show that the magnetic field, which interacts with the solar wind, is reduced when the magnetic orientation is aligned between Sun and Earth – opposite to the previous hypothesis. During forthcoming solar storms, this could result in blackouts and disruptions in artificial satellites.

Magnetic Field Reversals

Although generally Earth's field is approximately dipolar, with an axis that is nearly aligned with the rotational axis, occasionally the North and South geomagnetic poles trade places. Evidence for these *geomagnetic reversals* can be found in basalts, sediment cores taken from the ocean floors, and seafloor magnetic anomalies. Reversals occur nearly randomly in time, with intervals between reversals ranging from less than 0.1 million years to as much as 50 million years. The most recent geomagnetic reversal, called the Brunhes–Matuyama reversal, occurred about 780,000 years ago. A related phenomenon, a geomagnetic excursion, amounts to an incomplete reversal, with no change in polarity. The Laschamp event is an example of an excursion, it having occurred during the last ice age (41,000 years ago).

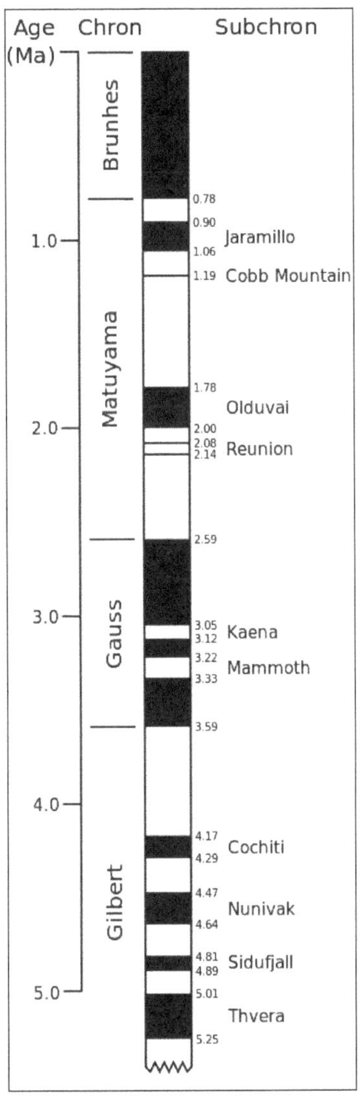

Geomagnetic polarity during the late Cenozoic Era. Dark areas denote periods where the polarity matches today's polarity, light areas denote periods where that polarity is reversed.

The past magnetic field is recorded mostly by strongly magnetic minerals, particularly iron oxides such as magnetite, that can carry a permanent magnetic moment. This remanent magnetization, or remanence, can be acquired in more than one way. In lava flows, the direction of the field is "frozen" in small minerals as they cool, giving rise to a thermoremanent magnetization. In sediments, the orientation of magnetic particles acquires a slight bias towards the magnetic field as they are deposited on an ocean floor or lake bottom. This is called detrital remanent magnetization.

Thermoremanent magnetization is the main source of the magnetic anomalies around mid-ocean ridges. As the seafloor spreads, magma wells up from the mantle, cools to form new basaltic crust on both sides of the ridge, and is carried away from it by seafloor spreading. As it cools, it records the direction of the Earth's field. When the Earth's field reverses, new basalt records the reversed direction. The result is a series of stripes that are symmetric about the ridge. A ship towing a magnetometer on the surface of the ocean can detect these stripes and infer the age of the ocean floor below. This provides information on the rate at which seafloor has spread in the past.

Radiometric dating of lava flows has been used to establish a *geomagnetic polarity time scale*, part of which is shown in the image. This forms the basis of magnetostratigraphy, a geophysical correlation technique that can be used to date both sedimentary and volcanic sequences as well as the seafloor magnetic anomalies.

Studies of lava flows on Steens Mountain, Oregon, indicate that the magnetic field could have shifted at a rate of up to 6 degrees per day at some time in Earth's history, which significantly challenges the popular understanding of how the Earth's magnetic field works. This finding was later attributed to unusual rock magnetic properties of the lava flow under study, not rapid field change, by one of the original authors of the 1995 study.

Temporary dipole tilt variations that take the dipole axis across the equator and then back to the original polarity are known as *excursions*.

Physical Origin

Earth's Core and the Geodynamo

The Earth's magnetic field is believed to be generated by electric currents in the conductive iron alloys of its core, created by convection currents due to heat escaping from the core. However the process is complex, and computer models that reproduce some of its features have only been developed in the last few decades.

The Earth and most of the planets in the Solar System, as well as the Sun and other stars, all generate magnetic fields through the motion of electrically conducting fluids. The Earth's field originates in its core. This is a region of iron alloys extending to about 3400 km (the radius of the Earth is 6370 km). It is divided into a solid inner core, with a radius of 1220 km, and a liquid outer core. The motion of the liquid in the outer core is driven by heat flow from the inner core, which is about 6,000 K (5,730 °C; 10,340 °F), to the core-mantle boundary, which is about 3,800 K (3,530 °C; 6,380 °F). The heat is generated by potential energy released by heavier materials sinking toward the core (planetary differentiation, the iron catastrophe) as well as decay of radioactive elements in the interior. The pattern of flow is organized by the rotation of the Earth and the presence of the solid inner core.

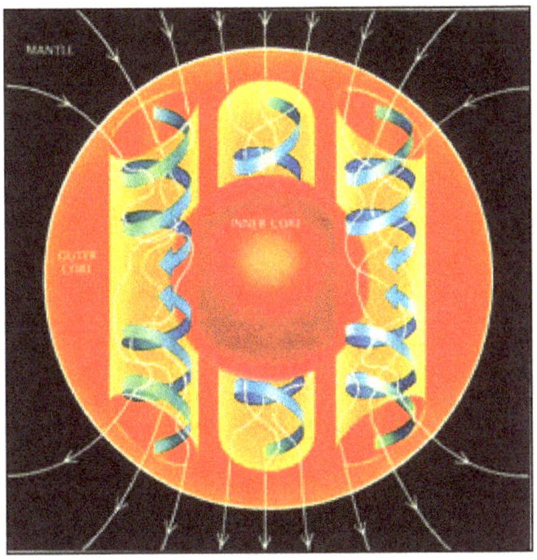

A schematic illustrating the relationship between motion of conducting fluid, organized into rolls by the Coriolis force, and the magnetic field the motion generates.

The mechanism by which the Earth generates a magnetic field is known as a dynamo. The magnetic field is generated by a feedback loop: current loops generate magnetic fields (Ampère's circuital law); a changing magnetic field generates an electric field (Faraday's law); and the electric and magnetic fields exert a force on the charges that are flowing in currents (the Lorentz force). These effects can be combined in a partial differential equation for the magnetic field called the *magnetic induction equation*:

$$\frac{\partial \mathrm{B}}{\partial t} = \eta \nabla^2 \mathrm{B} + \nabla \times (\mathrm{u} \times \mathrm{B}),$$

where u is the velocity of the fluid; B is the magnetic B-field; and $\eta = 1/\sigma\mu$ is the magnetic diffusivity, which is inversely proportional to the product of the electrical conductivity σ and the permeability μ . The term $\partial B/\partial t$ is the time derivative of the field; ∇^2 is the Laplace operator and $\nabla\times$ is the curl operator.

The first term on the right hand side of the induction equation is a diffusion term. In a stationary fluid, the magnetic field declines and any concentrations of field spread out. If the Earth's dynamo shut off, the dipole part would disappear in a few tens of thousands of years.

In a perfect conductor ($\sigma = \infty$), there would be no diffusion. By Lenz's law, any change in the magnetic field would be immediately opposed by currents, so the flux through a given volume of fluid could not change. As the fluid moved, the magnetic field would go with it. The theorem describing this effect is called the *frozen-in-field theorem*. Even in a fluid with a finite conductivity, new field is generated by stretching field lines as the fluid moves in ways that deform it. This process could go on generating new field indefinitely, were it not that as the magnetic field increases in strength, it resists fluid motion.

The motion of the fluid is sustained by convection, motion driven by buoyancy. The temperature increases towards the center of the Earth, and the higher temperature of the fluid lower down makes it buoyant. This buoyancy is enhanced by chemical separation: As the core cools, some of the molten iron solidifies and is plated to the inner core. In the process, lighter elements are left behind in the fluid, making it lighter. This is called *compositional convection*. A Coriolis effect, caused by the overall planetary rotation, tends to organize the flow into rolls aligned along the north-south polar axis.

A dynamo can amplify a magnetic field, but it needs a "seed" field to get it started. For the Earth, this could have been an external magnetic field. Early in its history the Sun went through a T-Tauri phase in which the solar wind would have had a magnetic field orders of magnitude larger than the present solar wind. However, much of the field may have been screened out by the Earth's mantle. An alternative source is currents in the core-mantle boundary driven by chemical reactions or variations in thermal or electric conductivity. Such effects may still provide a small bias that are part of the boundary conditions for the geodynamo.

The average magnetic field in the Earth's outer core was calculated to be 25 gausses, 50 times stronger than the field at the surface.

Numerical Models

Simulating the geodynamo requires numerically solving a set of nonlinear partial differential equations for the magnetohydrodynamics (MHD) of the Earth's interior. Simulation of the MHD equations is performed on a 3D grid of points and the fineness of the grid, which in part determines the realism of the solutions, is limited mainly by computer power. For decades, theorists were confined to creating *kinematic dynamo* computer models in which the fluid motion is chosen in advance and the effect on the magnetic field calculated. Kinematic dynamo theory was mainly a matter of trying different flow geometries and testing whether such geometries could sustain a dynamo.

The first *self-consistent* dynamo models, ones that determine both the fluid motions and the magnetic field, were developed by two groups in 1995, one in Japan and one in the United States. The latter received attention because it successfully reproduced some of the characteristics of the Earth's field, including geomagnetic reversals.

Currents in the Ionosphere and Magnetosphere

Electric currents induced in the ionosphere generate magnetic fields (ionospheric dynamo region). Such a field is always generated near where the atmosphere is closest to the Sun, causing daily alterations that can deflect surface magnetic fields by as much as one degree. Typical daily variations of field strength are about 25 nanoteslas (nT), with variations over a few seconds of typically around 1 nT (one part in 50,000).

Measurement and Analysis

Detection

The Earth's magnetic field strength was measured by Carl Friedrich Gauss in 1832 and has been repeatedly measured since then, showing a relative decay of about 10% over the last 150 years. The Magsat satellite and later satellites have used 3-axis vector magnetometers to probe the 3-D structure of the Earth's magnetic field. The later Ørsted satellite allowed a comparison indicating a dynamic geodynamo in action that appears to be giving rise to an alternate pole under the Atlantic Ocean west of South Africa.

Governments sometimes operate units that specialize in measurement of the Earth's magnetic field. These are geomagnetic observatories, typically part of a national Geological survey, for example the British Geological Survey's Eskdalemuir Observatory. Such observatories can measure and forecast magnetic conditions such as magnetic storms that sometimes affect communications, electric power, and other human activities.

The International Real-time Magnetic Observatory Network, with over 100 interlinked geomagnetic observatories around the world, has been recording the Earth's magnetic field since 1991.

The military determines local geomagnetic field characteristics, in order to detect *anomalies* in the natural background that might be caused by a significant metallic object such as a submerged submarine. Typically, these magnetic anomaly detectors are flown in aircraft like the UK's Nimrod or towed as an instrument or an array of instruments from surface ships.

Commercially, geophysical prospecting companies also use magnetic detectors to identify naturally occurring anomalies from ore bodies, such as the Kursk Magnetic Anomaly.

Crustal Magnetic Anomalies

Magnetometers detect minute deviations in the Earth's magnetic field caused by iron artifacts, kilns, some types of stone structures, and even ditches and middens in archaeological geophysics. Using magnetic instruments adapted from airborne magnetic anomaly detectors developed during World War II to detect submarines, the magnetic variations across the ocean floor have been mapped. Basalt — the iron-rich, volcanic rock making up the ocean floor — contains a strongly magnetic mineral (magnetite) and can locally distort compass readings. The distortion was recognized by Icelandic mariners as early as the late 18th century. More important, because the presence of magnetite gives the basalt measurable magnetic properties, these magnetic variations have provided another means to study the deep ocean floor. When newly formed rock cools, such magnetic materials record the Earth's magnetic field.

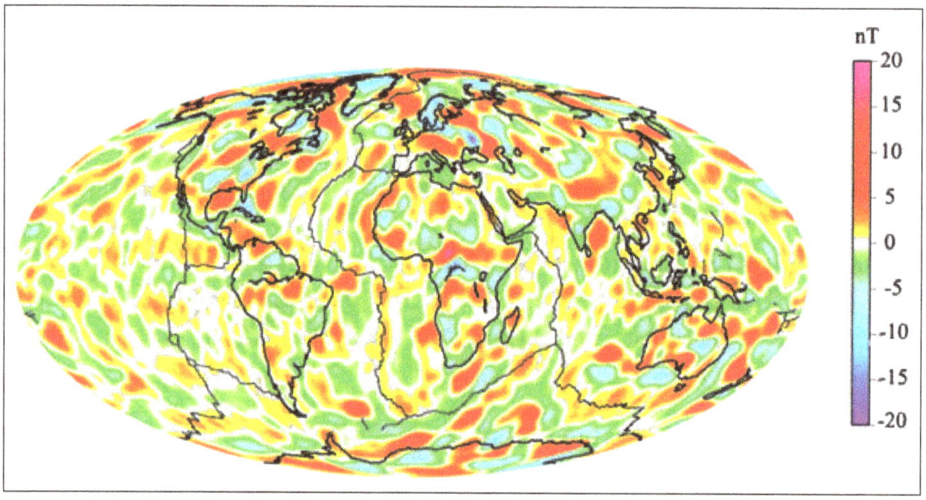

A model of short-wavelength features of Earth's magnetic field, attributed to lithospheric anomalies

Statistical Models

Each measurement of the magnetic field is at a particular place and time. If an accurate estimate of the field at some other place and time is needed, the measurements must be converted to a model and the model used to make predictions.

Spherical Harmonics

The most common way of analyzing the global variations in the Earth's magnetic field is to fit the measurements to a set of spherical harmonics. This was first done by Carl Friedrich Gauss. Spherical harmonics are functions that oscillate over the surface of a sphere. They are the product of two functions, one that depends on latitude and one on longitude. The function of longitude is zero along zero or more great circles passing through the North and South Poles; the number of such *nodal lines* is the absolute value of the *order* m. The function of latitude is zero along zero or more latitude circles; this plus the order is equal to the *degree* ℓ. Each harmonic is equivalent to a particular arrangement of magnetic charges at the center of the Earth. A *monopole* is an isolated magnetic charge, which has never been observed. A *dipole* is equivalent to two opposing charges brought close together and a *quadrupole* to two dipoles brought together.

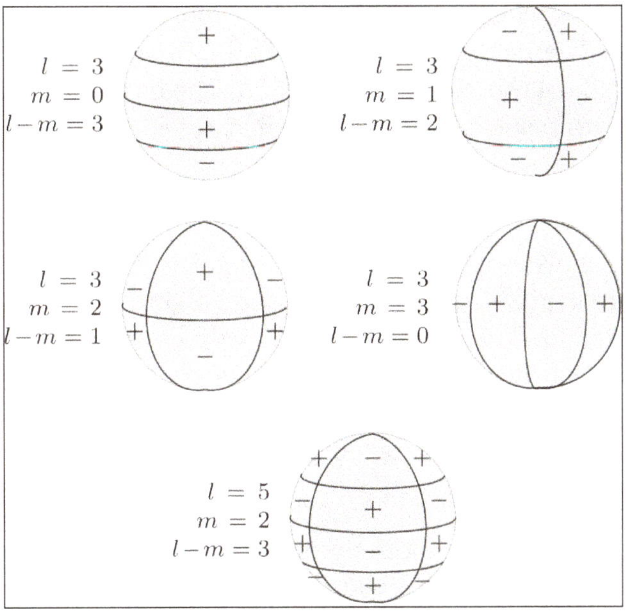

Schematic representation of spherical harmonics on a sphere and their nodal lines. $P_{\ell m}$ is equal to 0 along m great circles passing through the poles, and along ℓ-m circles of equal latitude. The function changes sign each ℓtime it crosses one of these lines.

Spherical harmonics can represent any scalar field (function of position) that satisfies certain properties. A magnetic field is a vector field, but if it is expressed in Cartesian components *X, Y, Z*, each component is the derivative of the same scalar function called the *magnetic potential*. Analyses of the Earth's magnetic field use a modified version of the usual spherical harmonics that differ by a multiplicative factor. A least-squares fit to the magnetic field measurements gives the Earth's field as the sum of spherical harmonics, each multiplied by the best-fitting *Gauss coefficient* g_m^ℓ or h_m^ℓ.

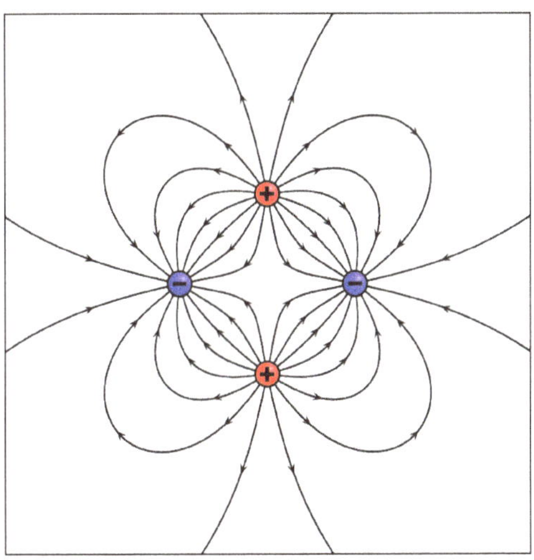

Example of a quadrupole field. This can also be constructed by moving two dipoles together.

The lowest-degree Gauss coefficient, g_0^0, gives the contribution of an isolated magnetic charge, so it is zero. The next three coefficients – g_1^0, g_1^1, and h_1^1 – determine the direction and magnitude of the dipole contribution. The best fitting dipole is tilted at an angle of about 10° with respect to the rotational axis,

Radial Dependence

Spherical harmonic analysis can be used to distinguish internal from external sources if measurements are available at more than one height (for example, ground observatories and satellites). In that case, each term with coefficient g_m^ℓ or h_m^ℓ can be split into two terms: one that decreases with radius as $1/r^{\ell+1}$ and one that *increases* with radius as r^ℓ. The increasing terms fit the external sources (currents in the ionosphere and magnetosphere). However, averaged over a few years the external contributions average to zero.

The remaining terms predict that the potential of a dipole source (ℓ=1) drops off as $1/r^2$. The magnetic field, being a derivative of the potential, drops off as $1/r^3$. Quadrupole terms drop off as $1/r^4$, and higher order terms drop off increasingly rapidly with the radius. The radius of the outer core is about half of the radius of the Earth. If the field at the core-mantle boundary is fit to spherical harmonics, the dipole part is smaller by a factor of about 8 at the surface, the quadrupole part by a factor of 16, and so on. Thus, only the components with large wavelengths can be noticeable at the surface. From a variety of arguments, it is usually assumed that only terms up to degree 14 or less have their origin in the core. These have wavelengths of about 2,000 kilometres (1,200 mi) or less. Smaller features are attributed to crustal anomalies.

Global Models

The International Association of Geomagnetism and Aeronomy maintains a standard global field model called the International Geomagnetic Reference Field. It is updated every five years. The 11th-generation model, IGRF11, was developed using data from satellites (Ørsted, CHAMP and SAC-C) and a world network of geomagnetic observatories. The spherical harmonic expansion was truncated at degree 10, with 120 coefficients, until 2000. Subsequent models are truncated at degree 13 (195 coefficients). Another global field model, called the World Magnetic Model, is produced jointly by the United States National Centers for Environmental Information and the British Geological Survey. This model truncates at degree 12 (168 coefficients) with an approximate spatial resolution of 3,000 kilometers. It is the model used by the United States Department of Defense, the Ministry of Defence, the United States Federal Aviation Administration (FAA), the North Atlantic Treaty Organization (NATO), and the International Hydrographic Office as well as in many civilian navigation systems. A third model, produced by the Goddard Space Flight Center (NASA and GSFC) and the Danish Space Research Institute, uses a "comprehensive modeling" approach that attempts to reconcile data with greatly varying temporal and spatial resolution from ground and satellite sources.

For users with higher accuracy needs, the United States National Centers for Environmental Information developed the Enhanced Magnetic Model (EMM), which extends to degree and order 790 and resolves magnetic anomalies down to a wavelength of 56 kilometers. It was compiled from satellite, marine, aeromagnetic and ground magnetic surveys. The latest version, EMM2017, includes data from The European Space Agency's Swarm satellite mission.

Magnetic Declination

Example of magnetic declination showing a compass needle with a "positive" (or "easterly") variation from geographic north. N_g is geographic or true north, N_m is magnetic north, and δ is magnetic declination.

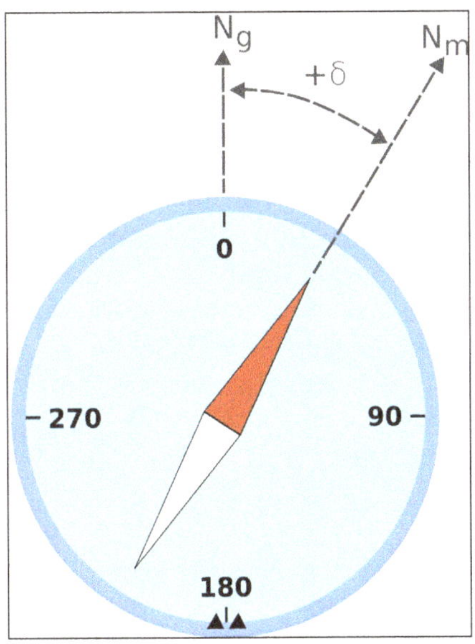

Magnetic declination, or magnetic variation, is the angle on the horizontal plane between magnetic north (the direction the north end of a magnetized compass needle points, corresponding to the direction of the Earth's magnetic field lines) and true north (the direction along a meridian towards the geographic North Pole). This angle varies depending on position on the Earth's surface and changes over time. Somewhat more formally, Bowditch defines variation as "the angle between the magnetic and geographic meridians at any place, expressed in degrees and minutes east or west to indicate the direction of magnetic north from true north. The angle between magnetic and grid meridians is called grid magnetic angle, grid variation, or grivation."

By convention, declination is positive when magnetic north is east of true north, and negative when it is to the west. *Isogonic lines* are lines on the Earth's surface along which the declination has the same constant value, and lines along which the declination is zero are called *agonic lines*. The lowercase letter δ is frequently used as the symbol for magnetic declination. The term *magnetic deviation* is sometimes used loosely to mean the same as magnetic declination, but more correctly it refers to the error in a compass reading induced by nearby metallic objects, such as iron on board a ship or aircraft.

Magnetic declination should not be confused with *magnetic inclination*, also known as magnetic dip, which is the angle that the Earth's magnetic field lines make with the downward side of the horizontal plane.

Declination Change over Time and Location

Magnetic declination varies both from place to place and with the passage of time. As a traveller cruises the east coast of the United States, for example, the declination varies from 16 degrees west in Maine, to 6 in Florida, to 0 degrees in Louisiana, to 4 degrees east. The declination at London, UK was one degree west, reducing to zero as of early 2020.

In most areas, the spatial variation reflects the irregularities of the flows deep in the Earth; in some areas, deposits of iron ore or magnetite in the Earth's crust may contribute strongly to the declination. Similarly, secular changes to these flows result in slow changes to the field strength and direction at the same point on the Earth.

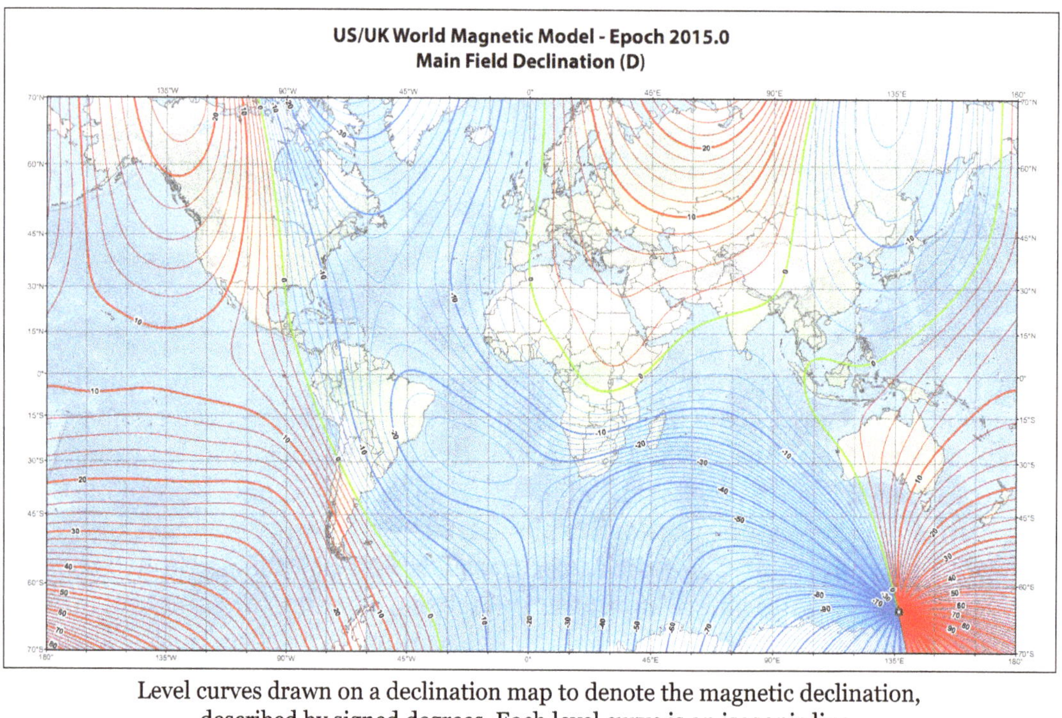

Level curves drawn on a declination map to denote the magnetic declination, described by signed degrees. Each level curve is an isogonic line.

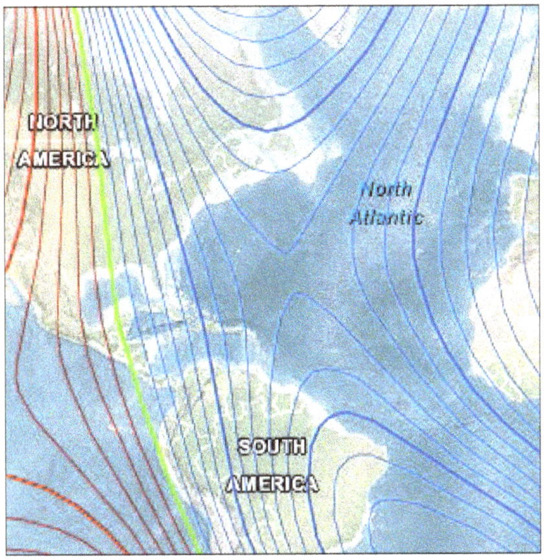

NIMA Magnetic Variation Map 2000.

The magnetic declination in a given area may (most likely will) change slowly over time, possibly as little as 2–2.5 degrees every hundred years or so, depending upon how far from the magnetic poles it is. For a location closer to the pole like Ivujivik, the declination may change by 1 degree every three years. This may be insignificant to most travellers, but can be important if using magnetic bearings from old charts or metes (directions) in old deeds for locating places with any precision.

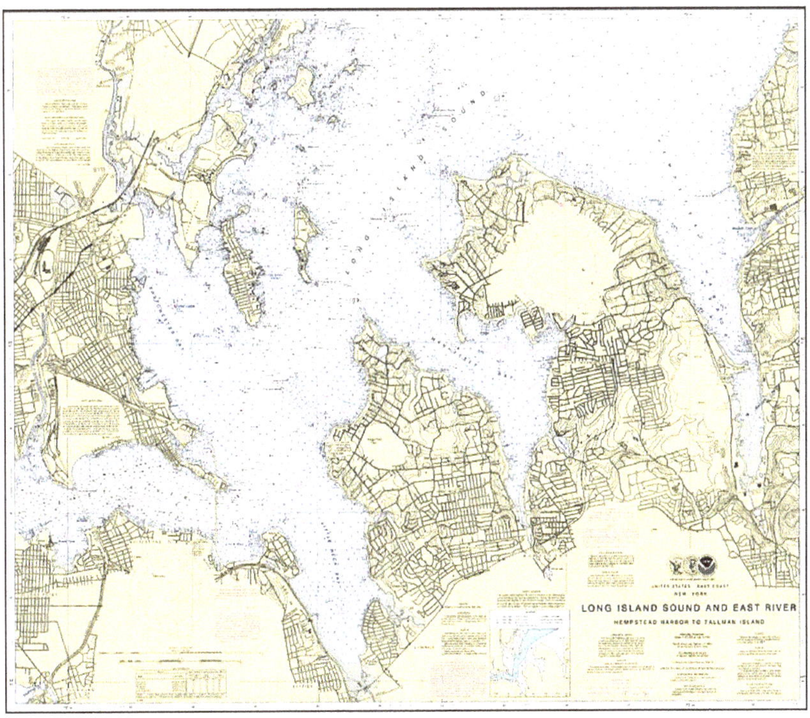

Western Long Island Sound, 2008.

As an example of how variation changes over time, see the two charts of the same area (western end of Long Island Sound), below, surveyed 124 years apart. The 1884 chart shows a variation of 8 degrees, 20 minutes West. The 2008 chart shows 13 degrees, 15 minutes West.

Determining Declination

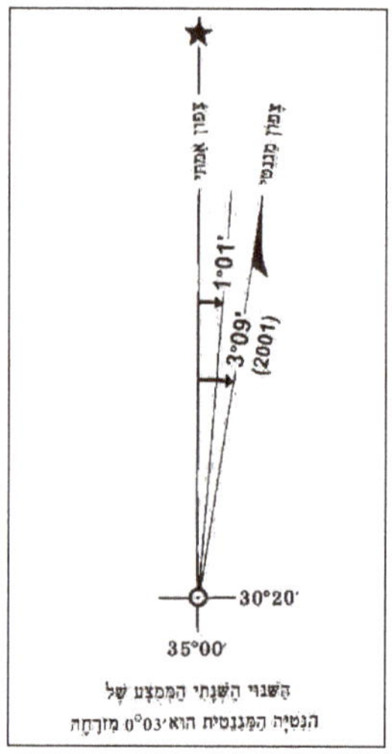

Magnetic declination indicated on an Israeli map. The arrows show true north, grid north and magnetic north, and the caption explains that the average yearly change in the magnetic declination is 0°03′ eastward.

Direct Measurement

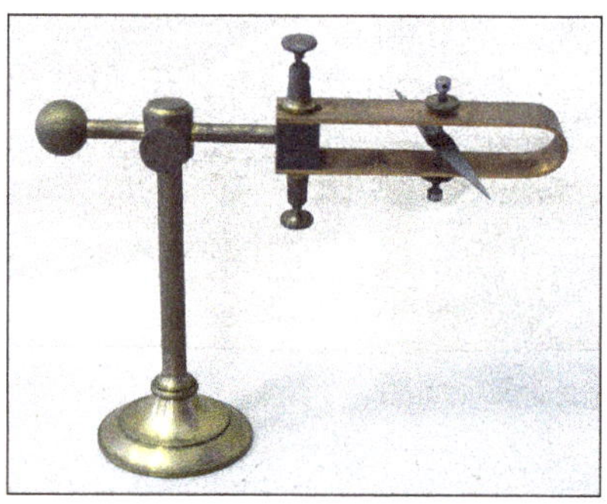

Antique declinometer.

The magnetic declination at any particular place can be measured directly by reference to the celestial poles—the points in the heavens around which the stars appear to revolve, which mark the direction of true north and true south. The instrument used to perform this measurement is known as a *declinometer*.

The approximate position of the north celestial pole is indicated by Polaris (the North Star). In the northern hemisphere, declination can therefore be approximately determined as the difference between the magnetic bearing and a visual bearing on Polaris. Polaris currently traces a circle 0.73° in radius around the north celestial pole, so this technique is accurate to within a degree. At high latitudes a plumb-bob is helpful to sight Polaris against a reference object close to the horizon, from which its bearing can be taken.

Determination from Maps and Models

A rough estimate of the local declination (within a few degrees) can be determined from a general isogonic chart of the world or a continent, such as those illustrated above. Isogonic lines are also shown on aeronautical and nautical charts.

Larger-scale local maps may indicate current local declination, often with the aid of a schematic diagram. Unless the area depicted is very small, declination may vary measurably over the extent of the map, so the data may be referred to a specific location on the map. The current rate and direction of change may also be shown, for example in arcminutes per year. The same diagram may show the angle of grid north (the direction of the map's north-south grid lines), which may differ from true north.

On the topographic maps of the U.S. Geological Survey, for example, a diagram shows the relationship between magnetic north in the area concerned (with an arrow marked "MN") and true north (a vertical line with a five-pointed star at its top), with a label near the angle between the MN arrow and the vertical line, stating the size of the declination and of that angle, in degrees, mils, or both.

A prediction of the current magnetic declination for a given location can be obtained online from a web page operated by the National Geophysical Data Center, a division of the National Oceanic and Atmospheric Administration of the United States. This model is built with all the information available to the map-makers at the start of the five-year period it is prepared for. It reflects a highly predictable rate of change, and is usually more accurate than a map which is likely months or years out of date and almost never less accurate.

Using the Declination

Adjustable Compasses

A magnetic compass points to magnetic north, not geographic north. Compasses of the style commonly used for hiking include a declination adjustment in the form of a bezel which swivels relative to the base plate. To establish a declination the bezel is rotated until the desired number of degrees plus or minus lie between the bezel's designation N (for North) and the direction indicated by the magnetic end of the needle (usually painted red). This allows the user to establish a true bearing for travel or orientation by aligning the embossed red indicator arrow on the base plate with a landmark or heading on a map. A compass thus adjusted can be said to be reading "true north" instead of magnetic north (as long as it remains within an area on the same isogonic line).

Adjustable compass set to a declination of 0° and a bearing of 307°.

In the image to the left, the bezel's N has been aligned with the direction indicated by the magnetic end of the compass needle, reflecting a magnetic declination of 0 degrees. The arrow on the base plate indicates a bearing of 307 degrees.

Non-adjustable Compasses

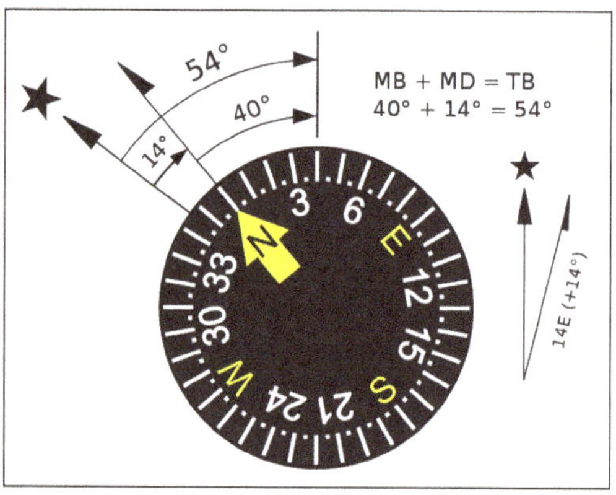

How to compensate for magnetic declination when reading a compass. In this example, the declination is 14°E (+14°), so the compass card points to a "north" 14 degrees to the East of true North. To obtain a true bearing, add 14 degrees to the bearing shown by the compass.

To work with both true and magnetic bearings, the user of a non-adjustable compass needs to make simple calculations that take into account the local magnetic declination. The example on the left shows how you would convert a magnetic bearing (one taken in the field using a non-adjustable compass) to a true bearing (one that you could plot on a map) by *adding* the magnetic declination. The declination in the example is 14°E (+14°). If, instead, the declination was 14°W (−14°), you would still "add" it to the magnetic bearing to obtain the true bearing: 40°+ (−14°) = 26°.

The opposite procedure is used in converting a true bearing to a magnetic bearing. With a local declination of 14°E, a true bearing (perhaps taken from a map) of 54° is converted to a magnetic bearing (for use in the field) by *subtracting* the declination: 54° − 14° = 40°. If, instead, the declination was 14°W (−14°), you would still "subtract" it from the true bearing to obtain the magnetic bearing: 54°- (−14°) = 68°.

Navigation

On aircraft or vessels there are three types of bearing: true, magnetic, and compass bearing. Compass error is divided into two parts, namely magnetic variation and magnetic deviation, the latter originating from magnetic properties of the vessel or aircraft. Variation and deviation are signed quantities. Positive (easterly) *variation* indicates that magnetic north is east of geographic north. Likewise, positive (easterly) *deviation* indicates that the compass needle is east of magnetic north.

Compass, magnetic and true bearings are related by:

$$T = M + V$$

$$M = C + D$$

The general equation relating compass and true bearings is:

$$T = C + D + V$$

Where:

- C is Compass bearing.
- M is Magnetic bearing.
- T is True bearing.
- V is magnetic Variation.
- D is compass Deviation.
- $V < 0, D < 0$ for Westerly Variation and Deviation.
- $V > 0,\ D > 0$ for Easterly Variation and Deviation.

For example, if the compass reads 32°, the local magnetic variation is −5.5° (i.e. West) and the deviation is 0.5° (i.e. East), the true bearing will be:

$$T = 32^\circ + (-5.5^\circ) + 0.5^\circ = 27^\circ$$

To calculate true bearing from compass bearing:

- Compass bearing + Deviation = Magnetic bearing.
- Magnetic bearing + Variation = True bearing.

To calculate compass bearing from true bearing:

- True bearing - Variation = Magnetic bearing.
- Magnetic bearing - Deviation = Compass bearing.

These rules are often combined with the mnemonic "West is Best, East is least"; that is to say, add W declinations when going from True bearings to Magnetic bearings, and subtract E ones.

Another simple way to remember which way to apply the correction for Continental USA is:

- For locations east of the agonic line (zero declination), roughly east of the Mississippi: The magnetic bearing is always bigger.
- For locations west of the agonic line (zero declination), roughly west of the Mississippi: The magnetic bearing is always smaller.

Common abbreviations are:

- TC = True course.
- V = Variation (of the Earth's magnetic field).
- MC = Magnetic course (what the course would be in the absence of local deviation).
- D = Deviation caused by magnetic material (mostly iron and steel) on the vessel.
- CC = Compass course.

Deviation

Magnetic deviation is the angle from a given magnetic bearing to the related bearing mark of the compass. Deviation is positive if a compass bearing mark (e.g., compass north) is right of the related magnetic bearing (e.g., magnetic north) and vice versa. For example, if the boat is aligned to magnetic north and the compass' north mark points 3° more east, deviation is +3°. Deviation varies for every compass in the same location and depends on such factors as the magnetic field of the vessel, wristwatches, etc. The value also varies depending on the orientation of the boat. Magnets and/or iron masses can correct for deviation, so that a particular compass accurately displays magnetic bearings.

More commonly, however, a correction card lists errors for the compass, which can then be compensated for arithmetically. Deviation must be added to compass bearing to obtain magnetic bearing.

Air Navigation

Magnetic declination has a very important influence on air navigation, since the most simple aircraft navigation instruments are designed to determine headings by locating magnetic north through the use of a compass or similar magnetic device.

Aviation sectionals (maps/charts) and databases used for air navigation are based on true north rather than magnetic north, and the constant and significant slight changes in the actual location of magnetic north and local irregularities in the planet's magnetic field require that charts and databases be updated at least twice each year to reflect the current magnetic variation correction from true north.

For example, near San Francisco, magnetic north is about 14.3 degrees east of true north, with the difference decreasing by about 6 minutes of arc per year. When plotting a course, most small aircraft pilots plot a trip using true north on a sectional, then convert the true north bearings to magnetic north for in-plane navigation using the magnetic compass. These bearings are then converted on a pre-flight plan by adding or subtracting the local variation displayed on a sectional.

Radionavigation aids located on the ground, such as VORs, are also checked and updated to keep them aligned with magnetic north to allow pilots to use their magnetic compasses for accurate and reliable in-plane navigation.

Runways are designated by a number between 01 and 36, which is generally one tenth of the magnetic azimuth of the runway's heading: a runway numbered 09 points east (90°), runway 18 is south (180°), runway 27 points west (270°) and runway 36 points to the north (360° rather than 0°). However, due to magnetic declination, changes in runway designators have to occur at times to keep their designation in line with the runway's magnetic heading. An exception is made for runways within the Northern Domestic Airspace of Canada; these are numbered relative to true north because proximity to the magnetic North Pole makes the magnetic declination large.

GPS systems used for air navigation can use magnetic north or true north. In order to make them more compatible with systems that depend on magnetic north, magnetic north is often chosen, at the pilot's preference. The GPS receiver natively reads in true north, but can elegantly calculate magnetic north based on its true position and data tables; the unit can then calculate the current location and direction of the north magnetic pole and (potentially) any local variations, if the GPS is set to use magnetic compass readings.

Magnetic Inclination

Magnetic dip, dip angle, or magnetic inclination is the angle made with the horizontal by the Earth's magnetic field lines. This angle varies at different points on the Earth's surface. Positive values of inclination indicate that the magnetic field of the Earth is pointing downward, into the Earth, at the point of measurement, and negative values indicate that it is pointing upward. The dip angle is in principle the angle made by the needle of a vertically held compass, though in practice ordinary compass needles may be weighted against dip or may be unable to move freely in the correct plane. The value can be measured more reliably with a special instrument typically known as a dip circle.

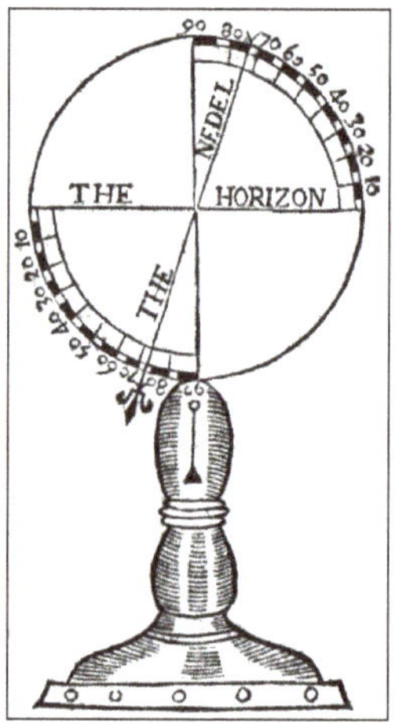

Illustration of magnetic dip.

Dip angle was discovered by the engineer Georg Hartmann in 1544. A method of measuring it with a dip circle was described by Robert Norman in England in 1581.

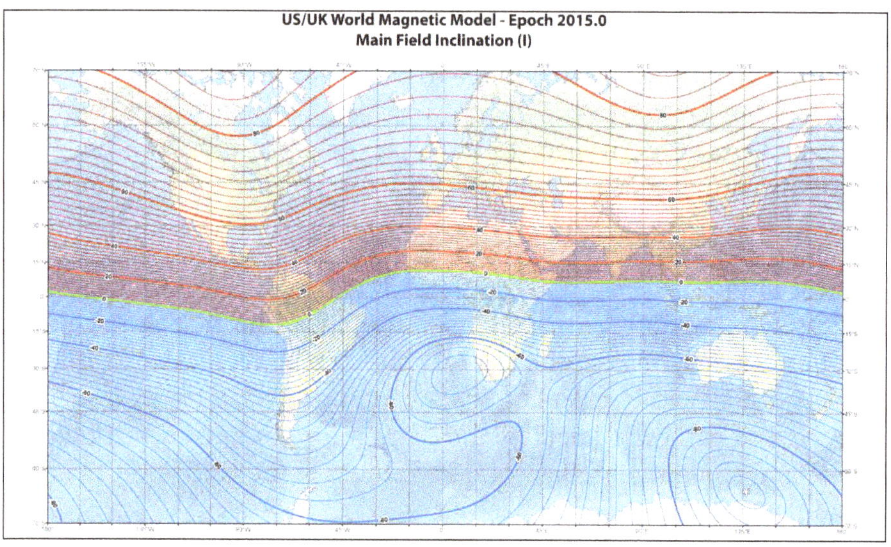

Isoclinic lines for the year.

Magnetic dip results from the tendency of a magnet to align itself with lines of magnetic field. As the Earth's magnetic field lines are not parallel to the surface, the north end of a compass needle will point downward in the northern hemisphere (positive dip) or upward in the southern hemisphere (negative dip). The range of dip is from -90 degrees (at the North Magnetic Pole) to +90 degrees (at the South Magnetic Pole). Contour lines along which the dip measured at the Earth's surface is equal are referred to as isoclinic lines. The locus of the points having zero dip is called the *magnetic equator* or aclinic line.

Practical Importance

The phenomenon is especially important in aviation, as it causes the airplane's compass to give erroneous readings during banked turns and airspeed changes. The latter errors occur because the compass card tilts on its mount when under acceleration.

Compass needles are often weighted during manufacture to compensate for magnetic dip, so that they will balance roughly horizontally. This balancing is latitude-dependent.

Dipole Model of the Earth's Magnetic Field

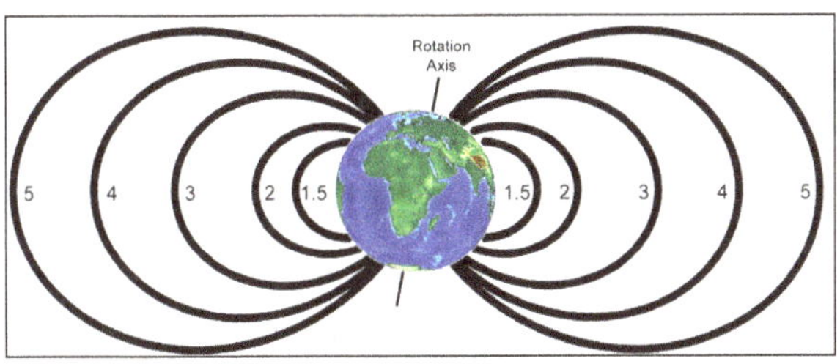

Plot showing field lines (which, in three dimensions would describe "shells") for L-values 1.5, 2, 3, 4 and 5 using a dipole model of the Earth's magnetic field.

The dipole model of the Earth's magnetic field is a first order approximation of the rather complex true Earth's magnetic field. Due to effects of the interplanetary magnetic field (IMF), and the solar wind, the dipole model is particularly inaccurate at high L-shells (e.g., above L=3), but may be a good approximation for lower L-shells. For more precise work, or for any work at higher L-shells, a more accurate model that incorporates solar effects, such as the Tsyganenko magnetic field model, is recommended.

Equations:

The following equations describe the dipole magnetic field.

First, define B_0 as the mean value of the magnetic field at the magnetic equator on the Earth's surface. Typically $B_0 = 3.12 \times 10^{-5}\ T$.

Then, the radial and azimuthal fields can be described as:

$$B_r = -2B_0\left(\frac{R_E}{r}\right)^3 \cos\theta$$

$$B_\theta = -B_0\left(\frac{R_E}{r}\right)^3 \sin\theta$$

$$|B| = B_0\left(\frac{R_E}{r}\right)^3 \sqrt{1+3\cos^2\theta}$$

where R_E is the mean radius of the Earth (approximately 6370 km), r is the radial distance from the center of the Earth (using the same units as used for R_E), and θ is the azimuth measured from the north magnetic pole (or geomagnetic pole).

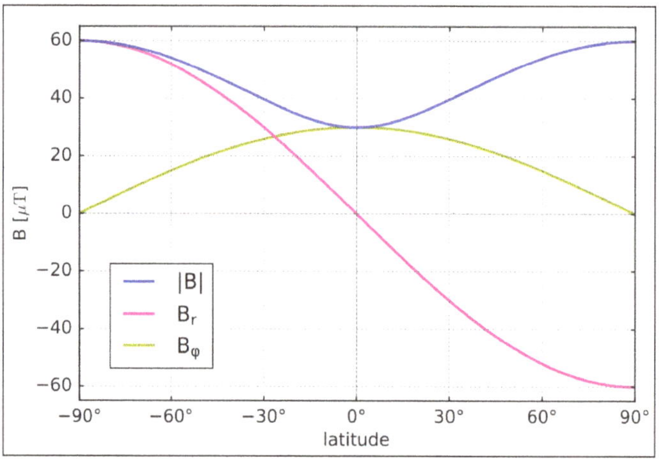

Magnetic field components vs. latitude.

It is sometimes more convenient to express the magnetic field in terms of magnetic latitude and distance in Earth radii. The magnetic latitude (MLAT), or geomagnetic latitude, λ is measured northwards from the equator (analogous to geographic latitude) and is related to θ by $\lambda = \pi/2 - \theta$. In this case, the radial and azimuthal components of the magnetic field (the latter still in the θ direction, measured from the axis of the north pole) are given by:

$$B_r = -\frac{2B_0}{R^3}\sin\lambda$$

$$B_\theta = \frac{B_0}{R^3}\cos\lambda$$

$$|B| = \frac{B_0}{R^3}\sqrt{1+3\sin^2\lambda}$$

where R in this case has units of Earth radii ($R = r/R_E$).

Invariant Latitude

Invariant latitude is a parameter that describes where a particular magnetic field line touches the surface of the Earth. It is given by:

$$\Lambda = \arccos\left(\sqrt{1/L}\right)$$

or,

$$L = 1/\cos^2(\Lambda)$$

where Λ is the invariant latitude and L is the L-shell describing the magnetic field line in question.

On the surface of the earth, the invariant latitude (Λ) is equal to the magnetic latitude (λ).

The Dynamo Effect

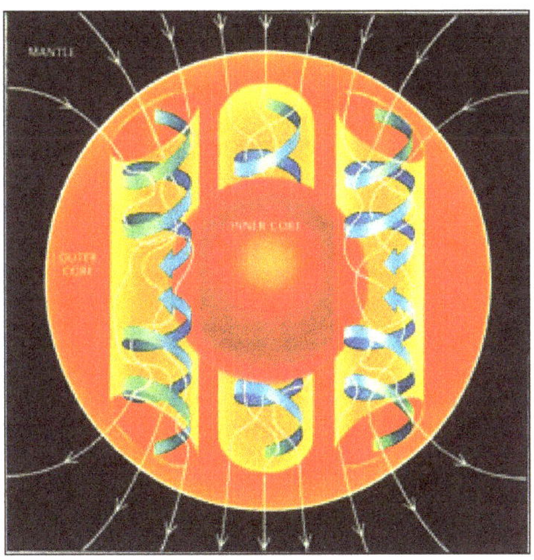

Illustration of the dynamo mechanism that creates the Earth's magnetic field: convection currents of fluid metal in the Earth's outer core, driven by heat flow from the inner core, organized into rolls by the Coriolis force, create circulating electric currents, which generate the magnetic field.

The dynamo theory proposes a mechanism by which a celestial body such as Earth or a star generates a magnetic field. The dynamo theory describes the process through which a rotating, convecting, and electrically conducting fluid can maintain a magnetic field over astronomical time scales. A dynamo is thought to be the source of the Earth's magnetic field and the magnetic fields of Mercury and the Jovian planets.

When William Gilbert published *de Magnete* in 1600, he concluded that the Earth is magnetic and proposed the first hypothesis for the origin of this magnetism: permanent magnetism such as that found in lodestone. In 1919, Joseph Larmor proposed that a dynamo might be generating the field. However, even after he advanced his hypothesis, some prominent scientists advanced alternative explanations. Einstein believed that there might be an asymmetry between the charges of the electron and proton so that the Earth's magnetic field would be produced by the entire Earth. The Nobel Prize winner Patrick Blackett did a series of experiments looking for a fundamental relation between angular momentum and magnetic moment, but found none.

Walter M. Elsasser, considered a "father" of the presently accepted dynamo theory as an explanation of the Earth's magnetism, proposed that this magnetic field resulted from electric currents induced in the fluid outer core of the Earth. He revealed the history of the Earth's magnetic field through pioneering the study of the magnetic orientation of minerals in rocks.

In order to maintain the magnetic field against ohmic decay (which would occur for the dipole field in 20,000 years), the outer core must be convecting. The convection is likely some combination of thermal and compositional convection. The mantle controls the rate at which heat is extracted from the core. Heat sources include gravitational energy released by the compression of the core, gravitational energy released by the rejection of light elements (probably sulfur, oxygen, or silicon) at the inner core boundary as it grows, latent heat of crystallization at the inner core boundary, and radioactivity of potassium, uranium and thorium.

At the dawn of the 21st century, numerical modeling of the Earth's magnetic field has not been successfully demonstrated, but appears to be in reach. Initial models are focused on field generation by convection in the planet's fluid outer core. It was possible to show the generation of a strong, Earth-like field when the model assumed a uniform core-surface temperature and exceptionally high viscosities for the core fluid. Computations which incorporated more realistic parameter values yielded magnetic fields that were less Earth-like, but also point the way to model refinements which may ultimately lead to an accurate analytic model. Slight variations in the core-surface temperature, in the range of a few millikelvins, result in significant increases in convective flow and produce more realistic magnetic fields.

Dynamo theory describes the process through which a rotating, convecting, and electrically conducting fluid acts to maintain a magnetic field. This theory is used to explain the presence of anomalously long-lived magnetic fields in astrophysical bodies. The conductive fluid in the geodynamo is liquid iron in the outer core, and in the solar dynamo is ionized gas at the tachocline. Dynamo theory of astrophysical bodies uses magnetohydrodynamic equations to investigate how the fluid can continuously regenerate the magnetic field.

It was once believed that the dipole, which comprises much of the Earth's magnetic field and is misaligned along the rotation axis by 11.3 degrees, was caused by permanent magnetization of the materials in the earth. This means that dynamo theory was originally used to explain the Sun's magnetic field in its relationship with that of the Earth. However, this hypothesis, which was initially proposed by Joseph Larmor in 1919, has been modified due to extensive studies of magnetic secular variation, paleomagnetism (including polarity reversals), seismology, and the solar system's abundance of elements. Also, the application of the theories of Carl Friedrich Gauss to magnetic observations showed that Earth's magnetic field had an internal, rather than external, origin.

There are three requisites for a dynamo to operate:

- An electrically conductive fluid medium.
- Kinetic energy provided by planetary rotation.
- An internal energy source to drive convective motions within the fluid.

In the case of the Earth, the magnetic field is induced and constantly maintained by the convection of liquid iron in the outer core. A requirement for the induction of field is a rotating fluid. Rotation in the outer core is supplied by the Coriolis effect caused by the rotation of the Earth. The Coriolis force tends to organize fluid motions and electric currents into columns aligned with the rotation axis. Induction or creation of magnetic field is described by the induction equation:

$$\frac{\partial \mathbf{B}}{\partial t} = \eta \nabla^2 \mathbf{B} + \nabla \times (\mathbf{u} \times \mathbf{B})$$

where u is velocity, B is magnetic field, t is time, and $\eta = 1/(\sigma\mu)$ is the magnetic diffusivity with σ electrical conductivity and μ permeability. The ratio of the second term on the right hand side to the first term gives the magnetic Reynolds number, a dimensionless ratio of advection of magnetic field to diffusion.

Tidal Heating Supporting a Dynamo

Tidal forces between celestial orbiting bodies cause friction that heats up their interiors. This is known as tidal heating, and it helps keep the interior in a liquid state. A liquid interior that can conduct electricity is required to produce a dynamo. Saturn's Enceladus and Jupiter's Io have enough tidal heating to liquify their inner cores, but they may not create a dynamo because they cannot conduct electricity. Mercury, despite its small size, has a magnetic field, because it has a conductive liquid core created by its iron composition and friction resulting from its highly elliptical orbit. It is theorized that the Moon once had a magnetic field, based on evidence from magnetized lunar rocks, due to its short-lived closer distance to Earth creating tidal heating. An orbit and rotation of a planet helps provide a liquid core, and supplements kinetic energy that supports a dynamo action.

Kinematic Dynamo Theory

In kinematic dynamo theory the velocity field is prescribed, instead of being a dynamic variable. This method cannot provide the time variable behavior of a fully nonlinear chaotic dynamo but is useful in studying how magnetic field strength varies with the flow structure and speed.

Using Maxwell's equations simultaneously with the curl of Ohm's law, one can derive what is basically the linear eigenvalue equation for magnetic fields (B) which can be done when assuming that the magnetic field is independent from the velocity field. One arrives at a critical magnetic Reynolds number above which the flow strength is sufficient to amplify the imposed magnetic field, and below which it decays.

The most functional feature of kinematic dynamo theory is that it can be used to test whether a velocity field is or is not capable of dynamo action. By applying a certain velocity field to a small magnetic field, it can be determined through observation whether the magnetic field tends to grow or not in reaction to the applied flow. If the magnetic field does grow, then the system is either capable of dynamo action or is a dynamo, but if the magnetic field does not grow, then it is simply referred to as non-dynamo.

The membrane paradigm is a way of looking at black holes that allows for the material near their surfaces to be expressed in the language of dynamo theory.

Spontaneous Breakdown of Topological Supersymmetry

Kinematic dynamo can be also viewed as the phenomenon of the spontaneous breakdown of the topological supersymmetry of the associated stochastic differential equation related to the flow of the background matter. Within supersymmetric theory of stochastics, this supersymmetry is an intrinsic property of all stochastic differential equations, its meaning is the preservation of the continuity of the phase space of the model by continuous time flows, and its spontaneous breakdown is the stochastic generalization of the concept of deterministic chaos. In other words, kinematic dynamo is a manifestation of the chaoticity of the underlying flow of the background matter.

Nonlinear Dynamo Theory

The kinematic approximation becomes invalid when the magnetic field becomes strong enough to affect the fluid motions. In that case the velocity field becomes affected by the Lorentz force, and so the induction equation is no longer linear in the magnetic field. In most cases this leads to a quenching of the amplitude of the dynamo. Such dynamos are sometimes also referred to as *hydromagnetic dynamos*. Virtually all dynamos in astrophysics and geophysics are hydromagnetic dynamos.

The main idea of the theory is that any small magnetic field existing in the outer core, creates currents in the moving fluid there due to Lorenz force. These currents create further magnetic field due to Ampere's law. With the fluid motion, the currents are carried in a way that the magnetic field gets stronger (as long as $u \cdot (J \times B)$ is negative). Thus a "seed" magnetic field can get stronger and stronger until it reaches some value that is related to existing non-magnetic forces.

Numerical models are used to simulate fully nonlinear dynamos. The following equations are used:

- The induction equation, presented above.
- Maxwell's equations for negligible electric field:

  $$\nabla \cdot \mathbf{B} = 0$$

  $$\nabla \times \mathbf{B} = \mu_0 \mathbf{J}$$

- The continuity equation for conservation of mass, for which the Boussinesq approximation is often used:

  $$\nabla \cdot \mathbf{u} = 0,$$

- The Navier-Stokes equation for conservation of momentum, again in the same approximation, with the magnetic force and gravitation force as the external forces:

  $$\frac{D\mathbf{u}}{Dt} = -\frac{1}{\rho_0}\nabla p + \nu \nabla^2 \mathbf{u} + \rho' \mathbf{g} + 2\Omega \times \mathbf{u} + \Omega \times \Omega \times \mathbf{R} + \frac{1}{\rho_0}\mathbf{J} \times \mathbf{B},$$

where ν is the kinematic viscosity, ρ_0 is the mean density and ρ' is the relative density perturbation that provides buoyancy (for thermal convection $\rho' = \alpha \Delta T$ where α is coefficient of thermal expansion), Ω is the rotation rate of the Earth, and $\mathbf{J}$ is the electric current density.

- A transport equation, usually of heat (sometimes of light element concentration):

  $$\frac{\partial T}{\partial t} = \kappa \nabla^2 T + \epsilon$$

where T is temperature, $\kappa = k / \rho c_p$ is the thermal diffusivity with k thermal conductivity, c_p heat capacity, and ρ density, and ϵ is an optional heat source. Often the pressure is the dynamic pressure, with the hydrostatic pressure and centripetal potential removed.

These equations are then non-dimensionalized, introducing the non-dimensional parameters,

$$Ra = \frac{g\alpha T D^3}{\nu\kappa}, E = \frac{\nu}{\Omega D^2}, Pr = \frac{\nu}{\kappa}, Pm = \frac{\nu}{\eta}$$

where *Ra* is the Rayleigh number, *E* the Ekman number, *Pr* and *Pm* the Prandtl and magnetic Prandtl number. Magnetic field scaling is often in Elsasser number units $B = (\rho\Omega / \sigma)^{1/2}$.

Energy Conversion between Magnetic and Kinetic Energy

The scalar product of the above form of Navier-Stokes equation with $\rho_0 \mathbf{u}$ gives the rate of increase of kinetic energy density, $(1/2)\rho_0 u^2$, on the left-hand side. The last term on the right-hand side is then $\mathbf{u}\cdot(\mathbf{J}\times\mathbf{B})$, the local contribution to the kinetic energy due to Lorentz force.

The scalar product of the induction equation with $(1/\mu_0)\mathbf{B}$ gives the rate of increase of the magnetic energy density, $(1/2\mu_0)B^2$, on the left-hand side. The last term on the right-hand side is then $(1/\mu_0)\mathbf{B}\cdot(\nabla\times(\mathbf{u}\times\mathbf{B}))$. Since the equation is volume-integrated, this term is equivalent up to a boundary term (and with the double use of the scalar triple product identity) to $-\mathbf{u}\cdot((1/\mu_0)(\nabla\times\mathbf{B})\times\mathbf{B}) = -\mathbf{u}\cdot(\mathbf{J}\times\mathbf{B})$ (where one of Maxwell's equations was used). This is the local contribution to the magnetic energy due to fluid motion.

Thus the term $-\mathbf{u}\cdot(\mathbf{J}\times\mathbf{B})$ is the rate of transformation of kinetic energy to magnetic energy. This has to be non-negative at least in part of the volume, for the dynamo to produce magnetic field.

Order of Magnitude of the Magnetic Field Created by Earth's Dynamo

The above formula for the rate of conversion of kinetic energy to magnetic energy, is equivalent to a rate of work done by a force of $\mathbf{J}\times\mathbf{B}$ on the outer core matter, whose velocity is $\mathbf{u}$. This work is the result of non-magnetic forces acting on the fluid.

Of those, the gravitational force and the centrifugal force are conservative and therefore have no overall contribution to fluid moving in closed loops. Ekman number (defined above), which is the ratio between the two remaining forces, namely the viscosity and Coriolis force, is very low inside Earth's outer core, because its viscosity is low (1.2-1.5 $\times 10^{-2}$ pascal-second) due to its liquidity.

Thus the main time-averaged contribution to the work is from Coriolis force, whose size is $-2\rho\mathbf{\Omega}\times\mathbf{u}$, though this quantity and $\mathbf{J}\times\mathbf{B}$ are related only indirectly and are not in general equal locally (thus they affect each other but not in the same place and time).

The current density *J* is itself the result of the magnetic field according to Ohm's law. Again, due to matter motion and current flow, this is not necessarily the field at the same place and time. However these relations can still be used to deduce orders of magnitude of the quantities in question.

In terms of order of magnitude, $JB \sim \rho\Omega u$ and $J \sim \sigma uB$, giving $\sigma uB^2 \sim \rho\Omega u$, or:

$$B \sim \sqrt{\frac{\rho\Omega}{\sigma}}$$

The exact ratio between both sides is the square root of Elsasser number.

The magnetic field direction cannot be inferred from this approximation (at least not its sign) as it appears squared, and is, indeed, sometimes reversed, though in general it lies on a similar axis to that of Ω .

For earth outer core, ρ is approximately 10^4 kg/m^3, Ω=2π/day = 7.3x10^{-5} seconds and σ is approximately $10^7 \Omega^{-1}m^{-1}$. This gives 2.7x10^{-4} Tesla.

The magnetic field of a magnetic dipole has an inverse cubic dependence in distance, so its order of magnitude at the earth surface can be approximated by multiplying the above result with $(R_{outer\ core}/R_{Earth})^3 = (2890/6370)^3 = 0.093$, giving 2.5x10^{-5} Tesla, not far from to the measured value of 3x10^{-5} Tesla at the equator.

Numerical Models

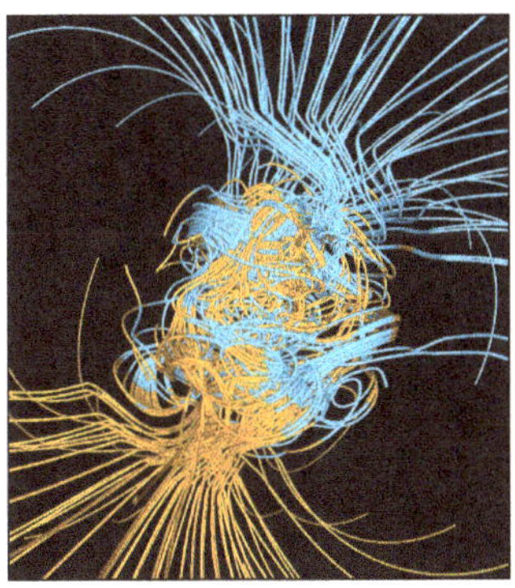

A visual representation of the Glatzmaier model before dipole reversal.

Broadly, models of the geodynamo attempt to produce magnetic fields consistent with observed data given certain conditions and equations.Implementing the magnetohydrodynamic equations successfully was of particular significance because they pushed dynamo models to self-consistency. Though geodynamo models are especially prevalent, dynamo models are not necessarily restricted to the geodynamo; solar and general dynamo models are also of interest. Studying dynamo models has utility in the field of geophysics as doing so can identify how various mechanisms form magnetic fields like those produced by astrophysical bodies like Earth and how they cause magnetic fields to exhibit certain features, such as pole reversals.

The equations used in numerical models of dynamo are highly complex. For decades, theorists were confined to two dimensional *kinematic dynamo* models, in which the fluid motion is chosen in advance and the effect on the magnetic field calculated. The progression from linear to nonlinear, three dimensional models of dynamo was largely hindered by the search for solutions to magnetohydrodynamic equations, which eliminate the need for many of the assumptions made in kinematic models and allow self-consistency.

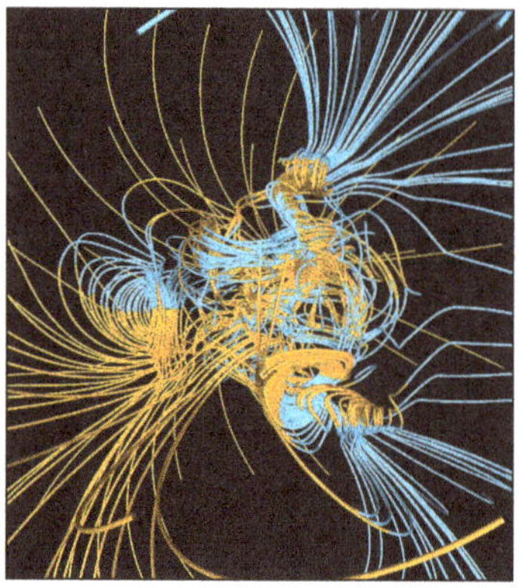

A visual representation of the Glatzmaier model during dipole reversal.

The first *self-consistent* dynamo models, ones that determine both the fluid motions and the magnetic field, were developed by two groups in 1995, one in Japan and one in the United States. The latter was made as a model with regards to the geodynamo and received significant attention because it successfully reproduced some of the characteristics of the Earth's field. Following this breakthrough, there was a large swell in development of reasonable, three dimensional dynamo models.

Though many self-consistent models now exist, there are significant differences among the models, both in the results they produce and the way they were developed. Given the complexity of developing a geodynamo model, there are many places where discrepancies can occur such as when making assumptions involving the mechanisms that provide energy for the dynamo, when choosing values for parameters used in equations, or when normalizing equations. In spite of the many differences that may occur, most models have shared features like clear axial dipoles. In many of these models, phenomena like secular variation and geomagnetic polarity reversals have also been successfully recreated.

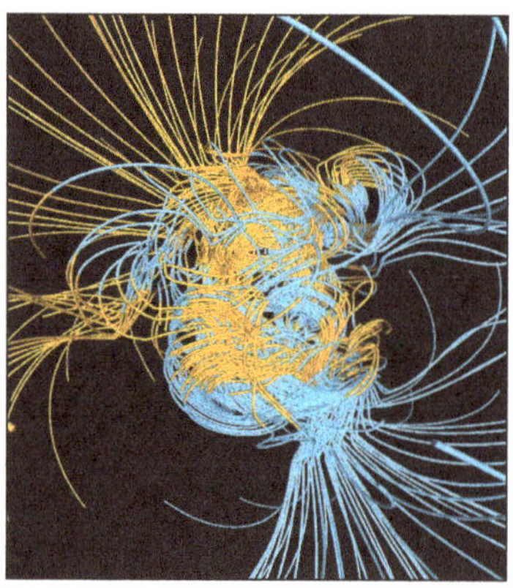

A visual representation of the Glatzmaier model after dipole reversal.

Observations

Many observations can be made from dynamo models. Models can be used to estimate how magnetic fields vary with time and can be compared to observed paleomagnetic data to find similarities between the model and the Earth. Due to the uncertainty of paleomagnetic observations, however, comparisons may not be entirely valid or useful. Simplified geodynamo models have shown relationships between the dynamo number (determined by variance in rotational rates in the outer core and mirror-asymmetric convection (e.g. when convection favors one direction in the north and the other in the south)) and magnetic pole reversals as well as found similarities between the geodynamo and the Sun's dynamo. In many models, it appears that magnetic fields have somewhat random magnitudes that follow a normal trend that average to zero. In addition to these observations, general observations about the mechanisms powering the geodynamo can be made based on how accurately the model reflects actual data collected from Earth.

Modern Modeling

The complexity of dynamo modelling is so great that models of the geodynamo are limited by the current power of supercomputers, particularly because calculating the Ekman and Rayleigh number of the outer core is extremely difficult and requires a vast number of computations.

Many improvements have been proposed in dynamo modelling since the self-consistent breakthrough in 1995. One suggestion in studying the complex magnetic field changes is applying spectral methods to simplify computations. Ultimately, until considerable improvements in computer power are made, the methods for computing realistic dynamo models will have to be made more efficient, so making improvements in methods for computing the model is of high importance for the advancement of numerical dynamo modelling.

Geomagnetic Reversal

Geomagnetic polarity during the last 5 million years. Dark areas denote periods where the polarity matches today's normal polarity; light areas denote periods where that polarity is reversed.

A geomagnetic reversal is a change in a planet's magnetic field such that the positions of magnetic north and magnetic south are interchanged. The Earth's field has alternated between periods of *normal* polarity, in which the predominant direction of the field was the same as the present direction, and *reverse* polarity, in which it was the opposite. These periods are called *chrons*.

Reversal occurrences are statistically random. There have been 183 reversals over the last 83 million years. The latest, the Brunhes–Matuyama reversal, occurred 780,000 years ago, with widely varying estimates of how quickly it happened. Other sources estimate that the time that it takes for a reversal to complete is on average around 7000 years for the four most recent reversals. Clement suggests that this duration is dependent on latitude, with shorter durations at low latitudes, and longer durations at mid and high latitudes. Although variable, the duration of a full reversal is typically between 2000 and 12000 years, which is one to two orders of magnitude less than the duration of magnetic chrons.

Although there have been periods in which the field reversed globally (such as the Laschamp excursion) for several hundred years, these events are classified as excursions rather than full geomagnetic reversals. Stable polarity chrons often show large, rapid directional excursions, which occur more often than reversals, and could be seen as failed reversals. During such an excursion, the field reverses in the liquid outer core, but not in the solid inner core. Diffusion in the liquid outer core is on timescales of 500 years or less, while that of the solid inner core is longer, around 3000 years.

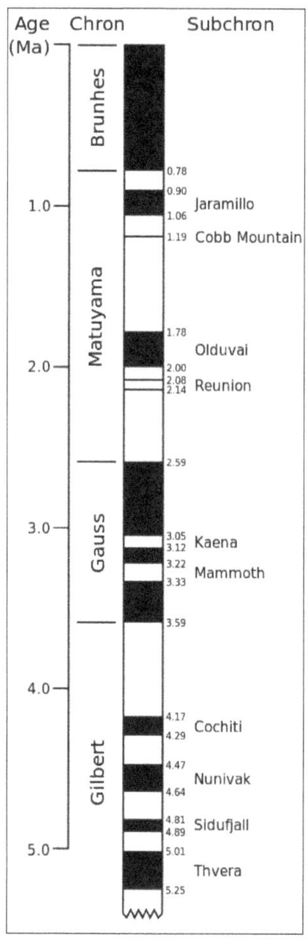

Observing Past Fields

Past field reversals can be and have been recorded in the "frozen" ferromagnetic (or, more accurately, ferrimagnetic) minerals of consolidated sedimentary deposits or cooled volcanic flows on land. The past record of geomagnetic reversals was first noticed by observing the magnetic stripe "anomalies" on the ocean floor. Lawrence W. Morley, Frederick John Vine and Drummond Hoyle Matthews made the connection to seafloor spreading in the Morley–Vine–Matthews hypothesis which soon led to the development of the theory of plate tectonics. The relatively constant rate at which the sea floor spreads results in substrate "stripes" from which past magnetic field polarity can be inferred from data gathered from towing a magnetometer along the sea floor.

Because no existing unsubducted sea floor (or sea floor thrust onto continental plates) is more than about 180 million years (Ma) old, other methods are necessary for detecting older reversals. Most sedimentary rocks incorporate tiny amounts of iron rich minerals, whose orientation is influenced by the ambient magnetic field at the time at which they formed. These rocks can preserve a record of the field if it is not later erased by chemical, physical or biological change.

Because the magnetic field is global, similar patterns of magnetic variations at different sites may be used to correlate age in different locations. In the past four decades much paleomagnetic data about seafloor ages (up to ~250 Ma) has been collected and is useful in estimating the age of geologic sections. Not an independent dating method, it depends on "absolute" age dating methods like radioisotopic systems to derive numeric ages. It has become especially useful to metamorphic and igneous geologists where index fossils are seldom available.

Geomagnetic Polarity Time Scale

Through analysis of seafloor magnetic anomalies and dating of reversal sequences on land, paleomagnetists have been developing a *Geomagnetic Polarity Time Scale* (GPTS). The current time scale contains 184 polarity intervals in the last 83 million years (and therefore 183 reversals).

Changing Frequency Over Time

The rate of reversals in the Earth's magnetic field has varied widely over time. 72 million years ago (Ma), the field reversed 5 times in a million years. In a 4-million-year period centered on 54 Ma, there were 10 reversals; at around 42 Ma, 17 reversals took place in the span of 3 million years. In a period of 3 million years centering on 24 Ma, 13 reversals occurred. No fewer than 51 reversals occurred in a 12-million-year period, centering on 15 million years ago. Two reversals occurred during a span of 50,000 years. These eras of frequent reversals have been counterbalanced by a few "superchrons" – long periods when no reversals took place.

Superchrons

A *superchron* is a polarity interval lasting at least 10 million years. There are two well-established superchrons, the Cretaceous Normal and the Kiaman. A third candidate, the Moyero, is more controversial. The Jurassic Quiet Zone in ocean magnetic anomalies was once thought to represent a superchron, but is now attributed to other causes.

The *Cretaceous Normal* (also called the *Cretaceous Superchron* or C34) lasted for almost 40 million years, from about 120 to 83 million years ago, including stages of the Cretaceous period from the Aptian through the Santonian. The frequency of magnetic reversals steadily decreased prior to the period, reaching its low point (no reversals) during the period. Between the Cretaceous Normal and the present, the frequency has generally increased slowly.

The *Kiaman Reverse Superchron* lasted from approximately the late Carboniferous to the late Permian, or for more than 50 million years, from around 312 to 262 million years ago. The magnetic field had reversed polarity. The name "Kiaman" derives from the Australian village of Kiama, where some of the first geological evidence of the superchron was found in 1925.

The Ordovician is suspected to have hosted another superchron, called the *Moyero Reverse Superchron*, lasting more than 20 million years (485 to 463 million years ago). Thus far, this possible superchron has only been found in the Moyero river section north of the polar circle in Siberia. Moreover, the best data from elsewhere in the world do not show evidence for this superchron.

Certain regions of ocean floor, older than 160 Ma, have low-amplitude magnetic anomalies that are hard to interpret. They are found off the east coast of North America, the northwest coast of Africa, and the western Pacific. They were once thought to represent a superchron called the *Jurassic Quiet Zone*, but magnetic anomalies are found on land during this period. The geomagnetic field is known to have low intensity between about 130 Ma and 170 Ma, and these sections of ocean floor are especially deep, causing the geomagnetic signal to be attenuated between the seabed and the surface.

Statistical Properties of Reversals

Several studies have analyzed the statistical properties of reversals in the hope of learning something about their underlying mechanism. The discriminating power of statistical tests is limited by the small number of polarity intervals. Nevertheless, some general features are well established. In particular, the pattern of reversals is random. There is no correlation between the lengths of polarity intervals. There is no preference for either normal or reversed polarity, and no statistical difference between the distributions of these polarities. This lack of bias is also a robust prediction of dynamo theory.

There is no *rate* of reversals, as they are statistically random. The randomness of the reversals is inconsistent with periodicity, but several authors have claimed to find periodicity. However, these results are probably artifacts of an analysis using sliding windows to attempt to determine reversal rates.

Most statistical models of reversals have analyzed them in terms of a Poisson process or other kinds of renewal process. A Poisson process would have, on average, a constant reversal rate, so it is common to use a non-stationary Poisson process. However, compared to a Poisson process, there is a reduced probability of reversal for tens of thousands of years after a reversal. This could be due to an inhibition in the underlying mechanism, or it could just mean that some shorter polarity intervals have been missed. A random reversal pattern with inhibition can be represented by a gamma process. In 2006, a team of physicists at the University of Calabria found that the reversals also conform to a Lévy distribution, which describes stochastic processes with long-ranging correlations between events in time. The data are also consistent with a deterministic, but chaotic, process.

Character of Transitions

Duration

Most estimates for the duration of a polarity transition are between 1,000 and 10,000 years, but some estimates are as quick as a human lifetime. Studies of 16.7-million-year-old lava flows on Steens Mountain, Oregon, indicate that the Earth's magnetic field is capable of shifting at a rate of up to 6 degrees per day. This was initially met with skepticism from paleomagnetists. Even if changes occur that quickly in the core, the mantle, which is a semiconductor, is thought to remove variations with periods less than a few months. A variety of possible rock magnetic mechanisms were proposed that would lead to a false signal. However, paleomagnetic studies of other sections from the same region (the Oregon Plateau flood basalts) give consistent results. It appears that the reversed-to-normal polarity transition that marks the end of Chron C5Cr (16.7 million years ago) contains a series of reversals and excursions. In addition, geologists Scott Bogue of Occidental College and Jonathan Glen of the US Geological Survey, sampling lava flows in Battle Mountain, Nevada, found evidence for a brief, several-year-long interval during a reversal when the field direction changed by over 50 degrees.

Magnetic Field

The magnetic field will not vanish completely, but many poles might form chaotically in different places during reversal, until it stabilizes again.

Causes

The magnetic field of the Earth, and of other planets that have magnetic fields, is generated by dynamo action in which convection of molten iron in the planetary core generates electric currents which in turn give rise to magnetic fields. In simulations of planetary dynamos, reversals often emerge spontaneously from the underlying dynamics. For example, Gary Glatzmaier and collaborator Paul Roberts of UCLA ran a numerical model of the coupling between electromagnetism and fluid dynamics in the Earth's interior. Their simulation reproduced key features of the magnetic field over more than 40,000 years of simulated time and the computer-generated field reversed itself. Global field reversals at irregular intervals have also been observed in the laboratory liquid metal experiment "VKS2".

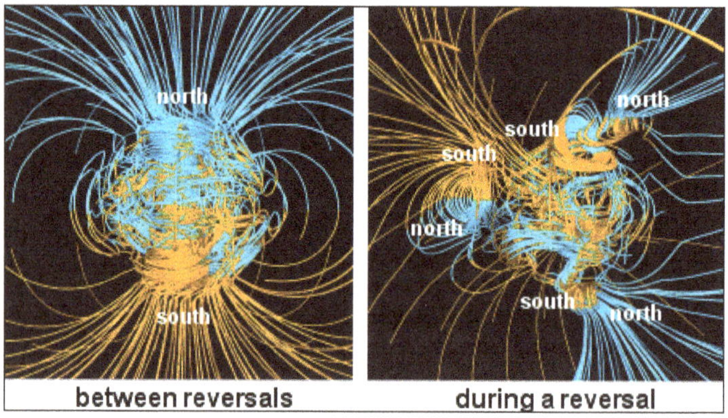

NASA computer simulation using the model of Glatzmaier and Roberts. The tubes represent magnetic field lines, blue when the field points towards the center and yellow when away. The rotation axis of the Earth is centered and vertical. The dense clusters of lines are within the Earth's core.

In some simulations, this leads to an instability in which the magnetic field spontaneously flips over into the opposite orientation. This scenario is supported by observations of the solar magnetic field, which undergoes spontaneous reversals every 9–12 years. However, with the Sun it is observed that the solar magnetic intensity greatly increases during a reversal, whereas reversals on Earth seem to occur during periods of low field strength.

Hypothesized Triggers

geomagnetic reversals are not spontaneous processes but rather are triggered by external events that directly disrupt the flow in the Earth's core. Proposals include impact events or internal events such as the arrival of continental slabs carried down into the mantle by the action of plate tectonics at subduction zones or the initiation of new mantle plumes from the core-mantle boundary. Supporters of this hypothesis hold that any of these events could lead to a large scale disruption of the dynamo, effectively turning off the geomagnetic field. Because the magnetic field is stable in either the present North-South orientation or a reversed orientation, they propose that when the field recovers from such a disruption it spontaneously chooses one state or the other, such that half the recoveries become reversals. However, the proposed mechanism does not appear to work in a quantitative model, and the evidence from stratigraphy for a correlation between reversals and impact events is weak. There is no evidence for a reversal connected with the impact event that caused the Cretaceous–Paleogene extinction event.

Effects on Biosphere

Shortly after the first geomagnetic polarity time scales were produced, scientists began exploring the possibility that reversals could be linked to extinctions. Most such proposals rest on the assumption that the Earth's magnetic field would be much weaker during reversals. Possibly the first such hypothesis was that high-energy particles trapped in the Van Allen radiation belt could be liberated and bombard the Earth. Detailed calculations confirm that if the Earth's dipole field disappeared entirely (leaving the quadrupole and higher components), most of the atmosphere would become accessible to high-energy particles, but would act as a barrier to them, and cosmic ray collisions would produce secondary radiation of beryllium-10 or chlorine-36. Greenland ice cores showed a peak of beryllium-10 during a brief complete reversal 41,000 years ago, which led to the magnetic field strength dropping to an estimated 5% of normal during the reversal. There is evidence that this occurs both during secular variation and during reversals.

Another hypothesis by McCormac and Evans assumes that the Earth's field disappears entirely during reversals. They argue that the atmosphere of Mars may have been eroded away by the solar wind because it had no magnetic field to protect it. They predict that ions would be stripped away from Earth's atmosphere above 100 km. However, paleointensity measurements show that the magnetic field has not disappeared during reversals. Based on paleointensity data for the last 800,000 years, the magnetopause is still estimated to have been at about three Earth radii during the Brunhes-Matuyama reversal. Even if the internal magnetic field did disappear, the solar wind can induce a magnetic field in the Earth's ionosphere sufficient to shield the surface from energetic particles.

Hypotheses have also advanced toward linking reversals to mass extinctions. Many such arguments were based on an apparent periodicity in the rate of reversals, but more careful analyses show that the reversal record is not periodic. It may be, however, that the ends of superchrons have caused vigorous convection leading to widespread volcanism, and that the subsequent airborne ash caused extinctions.

Tests of correlations between extinctions and reversals are difficult for a number of reasons. Larger animals are too scarce in the fossil record for good statistics, so paleontologists have analyzed microfossil extinctions. Even microfossil data can be unreliable if there are hiatuses in the fossil record. It can appear that the extinction occurs at the end of a polarity interval when the rest of that polarity interval was simply eroded away. Statistical analysis shows no evidence for a correlation between reversals and extinctions.

Magnetic Anomaly

A magnetic anomaly is a local variation in the Earth's magnetic field resulting from variations in the chemistry or magnetism of the rocks. Mapping of variation over an area is valuable in detecting structures obscured by overlying material. The magnetic variation in successive bands of ocean floor parallel with mid-ocean ridges is important evidence supporting the theory of seafloor spreading, central to plate tectonics.

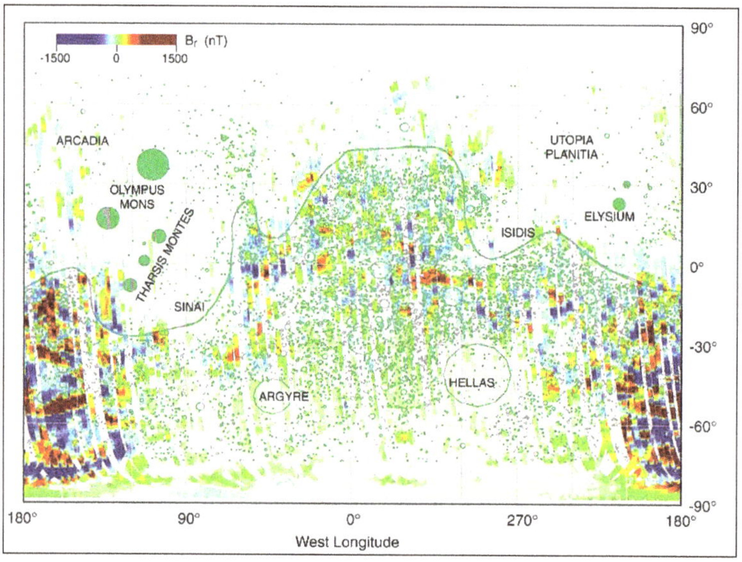

Satellite measurements of the radial magnetic field on Mars.

Measurement

Magnetic anomalies are generally a small fraction of the magnetic field. The total field ranges from 25,000 to 65,000 nanoteslas (nT). To measure anomalies, magnetometers need a sensitivity of 10 nT or less. There are three main types of magnetometer used to measure magnetic anomalies:

- The fluxgate magnetometer was developed during World War II to detect submarines. It measures the component along a particular axis of the sensor, so it needs to be oriented. On land, it is often oriented vertically, while in aircraft, ships and satellites it is usually oriented so the axis is in the direction of the field. It measures the magnetic field continuously, but drifts over time. One way to correct for drift is to take repeated measurements at the same place during the survey.

- The proton precession magnetometer measures the strength of the field but not its direction, so it does not need to be oriented. Each measurement takes a second or more. It is used in most ground surveys except for boreholes and high-resolution gradiometer surveys.

- Optically pumped magnetometers, which use alkali gases (most commonly rubidium and caesium) have high sample rates and sensitivities of 0.001 nT or less, but are more expensive than the other types of magnetometers. They are used on satellites and in most aeromagnetic surveys.

Data Acquisition

Ground-based

In ground-based surveys, measurements are made at a series of stations, typically 15 to 60 m apart. Usually a proton precession magnetometer is used and it is often mounted on a pole. Raising the magnetometer reduces the influence of small ferrous objects that were discarded by humans. To further reduce unwanted signals, they do not carry objects such as keys, knives or compasses. In addition, objects such as motor vehicles, railway lines, and barbed wire fences are avoided. If some such contaminant is overlooked, it often shows up as a sharp spike in the anomaly, so such features are treated with suspicion. The main application for ground-based surveys is detailed search for minerals.

Aeromagnetic

Airborne magnetic surveys are often used in oil surveys to provide preliminary information for seismic surveys. In some countries such as Canada, government agencies have made systematic surveys of large areas. The survey generally involves making a series of parallel runs at a constant height and intervals of anywhere from a hundred meters to several kilometers. These are crossed by occasional tie lines, perpendicular to the main survey, to check for errors. The plane is a source of magnetism, so sensors are either mounted on a boom or towed behind on a cable. Aeromagnetic surveys have a lower spatial resolution of ground surveys, but this can be an advantage for a regional survey of deeper rocks.

Shipborne

In shipborne surveys, a magnetometer is towed a few hundred meters behind a ship in a device called a *fish*. The sensor is kept at a constant depth of about 15 m. Otherwise, the procedure is similar to that used in aeromagnetic surveys.

Spacecraft

The Sputnik 3 spacecraft in 1958 was the first to carry a magnetometer. In the fall of 1979, Magsat was launched and jointly operated by NASA and USGS until the spring of 1980. It had a caesium vapor scalar magnetometer and a fluxgate vector magnetometer. CHAMP, a German satellite, made precise gravity and magnetic measurements from 2001 to 2010. A Danish satellite, Ørsted, was launched in 1999 and is still in operation, while the Swarm mission of the European Space Agency involves a "constellation" of three satellites that were launched in November, 2013.

Data Reduction

There are two main corrections that are needed for magnetic measurements. The first is remove short-term variations in the field from external sources. There are *diurnal variations* that have a period of 24 hours and magnitudes of up to 30 nT, probably from the action of the solar wind on the ionosphere. In addition, magnetic storms can have peak magnitudes of 1000 nT and can last for several days. Their contribution can be measured by returning to a base station repeatedly or by having another magnetometer that periodically measures the field at a fixed location.

The anomaly is the local contribution to the magnetic field, so the main geomagnetic field must be subtracted from it. Usually the International Geomagnetic Reference Field is used for this purpose. This is a large-scale, time-averaged mathematical model of the Earth's field based on measurements from satellites, magnetic observatories and other surveys. Some corrections that are needed for gravity anomalies are less important for magnetic anomalies. For example, the vertical gradient of the magnetic field is 0.03 nT/m or less, so an elevation correction is generally not needed.

Interpretation

The magnetization in the surveyed rock is a vector sum of induced and remanent magnetization: $\mathbf{M} = \mathbf{M}_i + \mathbf{M}_r$.

The induced magnetization of many minerals is the product of the ambient magnetic field and their magnetic susceptibility χ:

$$\mathbf{M}_i = \chi \mathbf{H}.$$

Minerals that are diamagnetic or paramagnetic only have an induced magnetization. Ferromagnetic minerals such as magnetite also can carry a remanent magnetization or remanence. This remanence can last for millions of years, so it may be in a completely different direction from the present Earth's field. If a remanence is present, it is difficult to separate from the induced magnetization unless samples of the rock are measured. The ratio of the magnitudes, $Q = M_r/M_i$, is called the Koenigsberger ratio.

Applications

Ocean Floor Stripes

Magnetic surveys over thc oceans have revealed a characteristic pattern of anomalies around mid-ocean ridges. They involve a series of positive and negative anomalies in the intensity of the magnetic field, forming stripes running parallel to each ridge. They are often symmetric about the axis of the ridge. The stripes are generally tens of kilometers wide, and the anomalies are a few hundred nanoteslas. The source of these anomalies is primarily permanent magnetization carried by titanomagnetite minerals in basalt and gabbros. They are magnetized when ocean crust is formed at the ridge. As magma rises to the surface and cools, the rock acquires a thermoremanent magnetization in the direction of the field. Then the rock is carried away from the ridge by the motions of the tectonic plates. Every few hundred thousand years, the direction of the magnetic field reverses. Thus, the pattern of stripes is a global phenomenon and can be used to calculate the velocity of seafloor spreading.

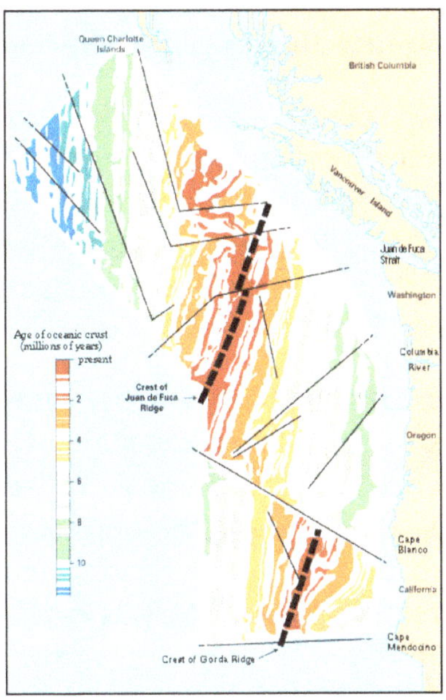

Magnetic anomalies around the juan de fuca and gorda ridges, off the west coast of north america, color-coded by age.

Variations of Earth's Magnetic Field

Short-term Variations

The geomagnetic field changes on time scales from milliseconds to millions of years. Shorter time scales mostly arise from currents in the ionosphere and magnetosphere, and some changes can be traced to geomagnetic storms or daily variations in currents. Changes over time scales of a year or more mostly reflect changes in the Earth's interior, particularly the iron-rich core.

Frequently, the Earth's magnetosphere is hit by solar flares causing geomagnetic storms, provoking displays of auroras.

Secular Variations

Changes in Earth's magnetic field on a time scale of a year or more are referred to as secular variation. Magnetic declination is observed to vary by tens of degrees over hundreds of years. Even the intensity of the dipole changes over time. Over the last two centuries the dipole strength has been decreasing at a rate of about 6.3% per century. At this rate of decrease, the field would reach zero in about 1600 years. However, this strength is about average for the last 7 thousand years, and the current rate of change is not unusual.

A prominent feature in the non-dipolar part of the secular variation is a westward drift. It has been noticed that the largr-scale anomalies of the geomagnetic field move westwards in a latitudinal direction at a rate of about 0.2 degrees per year. This drift is not the same everywhere and has varied over time. The globally averaged drift has been westward since about 1400 AD but eastward between about 1000 AD and 1400 AD.

Changes that predate magnetic observatories are recorded in archaeological and geological materials. Such changes are referred to as paleomagnetic secular variation or paleosecular variation (PSV). The records typically include long periods of small change with occasional large changes reflecting geomagnetic excursions and geomagnetic reversals.

Magnetic field reversals: Although the Earth's field is generally well approximated by a magnetic dipole with its axis near the rotational axis, there are occasional dramatic events where the North and South geomagnetic poles trade places. These events are called geomagnetic reversals. Evidence for these events can be found worldwide in basalts, sediment cores taken from the ocean floors, and seafloor magnetic anomalies. Reversals occur at apparently random intervals ranging from less than 0.1 million years to as much as 50 million years. The most recent such event, called the Brunhes–Matuyama reversal, occurred about 780,000 years ago.

Sources of Variation in the Steady Magnetic Field

Secular Variation of the Main Field

The main magnetic field of Earth, as observed at the surface, changes continuously with time. Changes of very short duration compared with geologic processes are called secular variation. Observations of declination made in London since 1540, for example, show that the direction of the field at that site has nearly completed a full cycle with a peak-to-peak amplitude of 30°. Other components of the field have been observed for a shorter length of time, but they also are exhibiting similar rapid change.

The characteristics of the secular variation are often represented by superimposing maps of the rate of change of a given field component on maps of the component itself. Such maps reveal that the world may be broken down into regions of continental scale in which a given component is either increasing or decreasing. Changes can be as large as 150 nanoteslas per year and persist for tens of years. If maps of secular variation from successively later times are examined, many features of the secular variation are found to be displaced westward with time.

The dominant component of the internal field is that of a centred dipole. It is useful to determine whether this component changes in the same way as the remainder of the field. Because the field of a dipole is so simple, it is more convenient to represent its change by its strength and orientation rather than by maps. Secular variation of the non-dipole components, however, are usually presented as maps. Such maps are similar to maps of secular variation of the entire field, indicating that most of the secular change is caused by the non-dipole components. On the average, the non-dipole components of the field appear to drift westward at an average rate of 0.18° per year. At this rate, drifting features circle Earth in only 2,000 years. Not all the non-dipole field exhibits drift. At least half of it appears fixed and variable only in intensity.

The dipole component also changes with time. Since 1850 its strength has decreased from about 8.5 × 1022 to about 8 × 1022 amperes per square metre. If this trend continues, the dipole component will vanish in another 2,000 years. The dipole component of Earth's field appears to be in the process of reversing. The best estimates are that the orientation of the dipole component appears to change with time. The dominant change is a westward drift of the azimuth of the dipole but at a rate much slower (0.08° per year) than the non-dipole component. The polar angles also may be increasing but even more slowly.

The origin of the secular variation is not known. Investigators suspect that it is a secondary effect of the dynamo mechanism that generates the main field. The short timescale of the variation implies that the source is in the outer region of the liquid core. If the source was deeper, the variation would be so attenuated by the electrical conductivity of the core that it would be undetectable at the surface.

The westward drift of magnetic anomalies evident in the secular variation should provide an important clue to the origin of the main field if only it can be interpreted. One model explains the drift by postulating that the outer portion of the liquid core is rotating slower than the more rigid mantle above. As a whole, Earth rotates eastward. If features within the core rotate more slowly than surface features, they will appear to move backward relative to the general rotation—i.e., westward. In this model the secular variation is caused by portions of eddies in the internal current system that rotate more slowly than the planet as a whole.

A more recent model for the westward drift posits that it is produced by hydromagnetic waves in the core. In this model the core rotates at the same rate as the outer mantle, but a wave propagates slowly around the outer portion of the core. Because waves in a conducting fluid distort the magnetic field frozen within it, they produce changes that can be observed at the surface. Since the characteristics of waves depend on the medium through which they propagate, it may be possible to infer properties of the outer core from surface observations.

Reversals of the Main Field

Earth's internal magnetic field has not always been oriented as it is today. The direction of the dipole component reverses, on an average, about every 300,000 to 1,000,000 years. This reversal is very sudden on a geologic timescale, apparently taking about 5,000 years. The time between reversals is highly variable, sometimes occurring in less than 40,000 years and at other times remaining steady for as long as 35,000,000 years. No regularities or periodicities have yet been discovered in the pattern of reversals. A long interval of one polarity may be followed by a short interval of opposite polarity.

Available data suggest that during a reversal the strength of the dipole component shrinks to zero while maintaining its orientation. It then grows again to its former strength but with opposite orientation. During the interval in which there is no dipole component, the non-dipole part of the field appears to persist.

During field reversals the outer portion of Earth's magnetic field is greatly altered. The absence of a dipole component would mean that the solar wind would approach much closer to Earth. Cosmic-ray particles that are normally deflected by Earth's field or are trapped in its outer portions would reach the surface of the planet. These particles might cause genetic damage in plant or animal communities, leading to the disappearance of one species and the appearance of another. Attempts have been made to establish whether there is evidence for such changes at the time of field reversals. Thus far the results remain inconclusive.

Evidence for the occurrence of magnetic reversals is unquestionable, however. Magnetic surveys made by ship across spreading centres in the middle of the oceans provide the best evidence. These data show that strips of oppositely magnetized ocean floor appear symmetrically about such features as the Mid-Atlantic Ridge. The explanation for these strips is that molten basalt flows out of the ridge and spreads away in both directions. As the basalt cools, it captures the orientation of the prevailing magnetic field and carries it along on the spreading seafloor. Basalt emerging from the ridge and cooling at later times captures the subsequent field orientation. The seafloor thus acts like a magnetic tape, capturing the alternating sequence of field orientations.

It should be noted that more information than the sense of the dipole component is captured in cooling rocks. Rocks formed at the magnetic equator, for example, contain a horizontal magnetization. Similarly, rocks formed at higher magnetic latitudes contain a field pointing up or down at an inclination that depends on latitude. The declination of the magnetization further reveals the direction to the magnetic pole at the time of the magnetization. Together these two angles can be used to infer the location of a virtual magnetic pole relative to the location of the sample.

Such a technique has been used to study the history of Earth's field at various locations. When virtual poles are determined from progressively older rocks, it is found that the virtual poles appear to wander with time. For many years it was thought that this "polar wandering" was a characteristic of Earth's magnetic field. Recent studies, however, prove instead that it is a result of continental drift. Magnetic poles have not moved significantly relative to the geographic poles, but rather the continents have. Thus, progressively older rocks were formed when continents were at different locations from where they are today.

Reversals of the main field must be caused by the dynamo mechanism that gives rise to the field in the first place. The timescale for the reversal is so rapid that it clearly cannot be caused by geologic processes. Furthermore, reversals cannot be caused by simple decay and reappearance of a preexisting field. The electrical conductivity of the core is too high to allow the field to decay on such a short timescale. In some way minor changes in the magnetic field configuration of the core must be amplified by thermal convection, causing the field to grow rapidly in the opposite direction. Models that simulate the main field have been shown to possess this property. The solutions to equations that describe the generation of the main field are unstable, and small changes can cause solutions of opposite sign to appear.

Variations in the Ionospheric Dynamo Current

The ionospheric dynamo is produced by movement of charged particles of the ionosphere across Earth's main field. This motion is driven by the tidal effects of the Sun and the Moon and by solar heating. The ionospheric dynamo is thus controlled by two parameters: the distribution of winds and the distribution of electrical conductivity in the ionosphere. These parameters are influenced by several factors, including the orbital parameters of Earth, Moon, and Sun; the solar cycle; solar flares; and solar eclipses. Changes in the position of the Sun and the Moon relative to Earth as a result of orbital motions cause variations in distance. This alters the strength of the tides and of solar heating, thereby changing ionospheric wind patterns. These changes are apparent as a seasonal modulation of the winds and hence of the strength of the current.

The second parameter that controls the dynamo current is the electrical conductivity of the ionosphere. Any process that alters ionospheric conductivity changes the current. On the dayside of Earth the dominant source of ionization is sunlight. The amount of ionization depends on the angle at which sunlight enters the atmosphere. Vertical incidence produces more ionization per unit volume than slant entry. For a given hemisphere, normal incidence occurs in summer. Thus, this effect also causes a strong seasonal modulation of the dynamo current.

The degree of atmospheric ionization also depends on the phase of the solar cycle. This 11-year cycle of sunspot activity produces variations in the amount of ultraviolet radiation emitted by the Sun. More sunspots lead to more ultraviolet radiation and increased ionospheric conductivity and hence stronger currents. On a shorter timescale solar flares emit X-rays that penetrate deeper in the atmosphere, temporarily ionizing the D region, the lowest layer of the ionosphere. Dynamo currents are then produced in this layer by whatever winds are present there.

A solar eclipse produces the opposite effect on ionospheric conductivity. The shadow of the Moon as it crosses the ionosphere decreases ionization. Recombination of ionospheric electrons and ions in the absence of light quickly reduces the conductivity. Because the effect is localized and of short duration, its effect on the overall dynamo current is slight.

Paleomagnetism

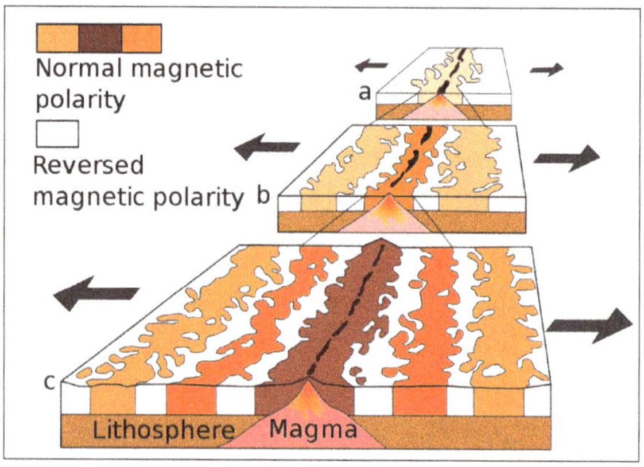

Magnetic stripes are the result of reversals of the Earth's field and seafloor spreading. New oceanic crust is magnetized as it forms and then it moves away from the ridge in both directions. The models show a ridge (a) about 5 million years ago (b) about 2 to 3 million years ago and (c) in the present.

Paleomagnetism is the study of the record of the Earth's magnetic field in rocks, sediment, or archeological materials. Certain minerals in rocks lock-in a record of the direction and intensity of the magnetic field when they form. This record provides information on the past behavior of Earth's magnetic field and the past location of tectonic plates. The record of geomagnetic reversals preserved in volcanic and sedimentary rock sequences (magnetostratigraphy) provides a time-scale that is used as a geochronologic tool. Geophysicists who specialize in paleomagnetism are called *paleomagnetists*.

Paleomagnetists led the revival of the continental drift hypothesis and its transformation into plate tectonics. Apparent polar wander paths provided the first clear geophysical evidence for continental drift, while marine magnetic anomalies did the same for seafloor spreading. Paleomagnetism continues to extend the history of plate tectonics back in time and are applied to the movement of continental fragments, or terranes.

Paleomagnetism relied heavily on new developments in rock magnetism, which in turn has provided the foundation for new applications of magnetism. These include biomagnetism, magnetic fabrics (used as strain indicators in rocks and soils), and environmental magnetism.

Fields

Paleomagnetism is studied on a number of scales:

- Secular variation studies look at small-scale changes in the direction and intensity of the Earth's magnetic field. The magnetic north pole is constantly shifting relative to the axis of rotation of the Earth. Magnetism is a vector and so magnetic field variation is made up of palaeodirectional measurements of magnetic declination and magnetic inclination and palaeointensity measurements.

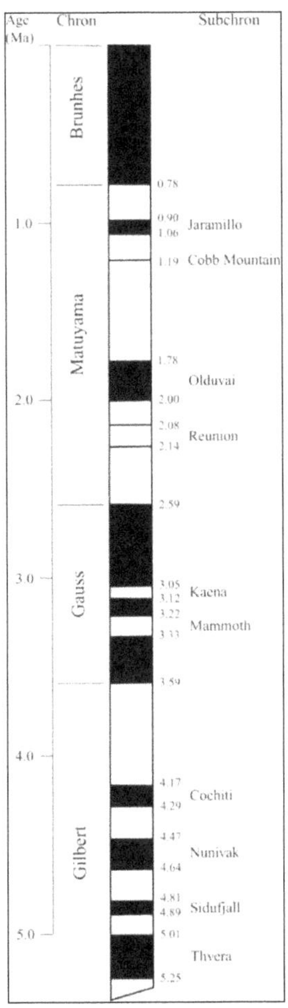

Earth's magnetic polarity reversals in last 5 million years. Dark regions represent normal polarity (same as present field); light regions represent reversed polarity.

- Magnetostratigraphy uses the polarity reversal history of the Earth's magnetic field recorded in rocks to determine the age of those rocks. Reversals have occurred at irregular intervals throughout Earth history. The age and pattern of these reversals is known from the study of sea floor spreading zones and the dating of volcanic rocks.

Principles of Remanent Magnetization

The study of paleomagnetism is possible because iron-bearing minerals such as magnetite may record past directions of the Earth's magnetic field. Magnetic signatures in rocks can be recorded by several different mechanisms.

Thermoremanent Magnetization

When an igneous rock cools, it acquires a thermoremanent magnetization (TRM) from the Earth's field. TRM can be much larger than it would be if exposed to the same field at room temperature. This remanence can also be very stable, lasting without significant change for millions of years. TRM is the main reason that paleomagnetists are able to deduce the direction and magnitude of the ancient Earth's field.

Paleomagnetism

Demagnetization

It has long been known that a TRM can be removed if it is heated above the Curie temperature T_C of the minerals carrying it. A TRM can also be partially demagnetized by heating up to some lower temperature T_1 and cooling back to room temperature. A common procedure in paleomagnetism is *stepwise demagnetization*, in which the sample is heated to a series of temperatures $T_1, T_2, \ldots$, cooling to room temperature and measuring the remaining remanence in between each heating step. The series of remanences can be plotted in a variety of ways, depending on the application.

Partial TRM

If a rock is later re-heated (as a result of burial, for example), part or all of the TRM can be replaced by a new remanence. If it is only part of the remanence, it is known as *partial thermoremanent magnetization (pTRM)*. Because numerous experiments have been done modeling different ways of acquiring remanence, pTRM can have other meanings. For example, it can also be acquired in the laboratory by cooling in zero field to a temperature T_1 (below the Curie temperature), applying a magnetic field and cooling to a temperature T_2, then cooling the rest of the way to room temperature in zero field.

Ideal TRM Behavior

Thellier Laws

The ideal TRM is one that can record the magnetic field in such a way that both its direction and intensity can be measured by some process in the lab. Thellier showed that this could be done if pTRM's satisfied four laws. Suppose that A and B are two non-overlapping temperature intervals. Suppose that M_A is a pTRM that is acquired by cooling the sample to room temperature, only switching the field H on while the temperature is in interval A; M_B has a similar definition. The *Thellier laws* are:

- Linearity: $M_A(H)$ and $M_B(H)$ are proportional to H when H is not much larger than the present Earth's field.
- Reciprocity: M_A can be removed by heating through temperature interval A, and M_B through B.
- Independence: M_A and M_B are independent.
- Additivity: If $M_{A\cup B}$ is acquired by turning the field on in both temperature intervals, $M_{A\cup B} = M_A + M_B$.

If these laws hold for any non-overlapping temperature intervals A and B, the sample satisfies the Thellier laws.

Simple Model for the Thellier Laws

Suppose that a sample has a lot of magnetic minerals, each of which has the following property: It is superparamagnetic until the temperature reaches a *blocking temperature* T_B that is independent of magnetic field for small fields. No irreversible changes occur at temperatures below T_B. If the resulting TRM is heated in zero field, it becomes superparamagnetic again at an *unblocking temperature* T_{UB} that is equal to T_B. Then it is easy to verify that reciprocity, independence and additivity hold. It only remains for linearity to be satisfied for all the Thellier laws to be obeyed.

Néel Model for Single-domain TRM

Louis Néel developed a physical model that showed how real magnetic minerals could have the above properties. It applies to particles that are single-domain, having a uniform magnetization that can only rotate as a unit.

Detrital Remanent Magnetization

In a completely different process, magnetic grains in sediments may align with the magnetic field during or soon after deposition; this is known as *detrital remanent magnetization* (DRM). If the magnetization is acquired as the grains are deposited, the result is a depositional detrital remanent magnetization (dDRM); if it is acquired soon after deposition, it is a post-depositional detrital remanent magnetization (pDRM).

Chemical Remanent Magnetization

In a third process, magnetic grains grow during chemical reactions, and record the direction of the magnetic field at the time of their formation. The field is said to be recorded by *chemical remanent magnetization* (CRM). A common form of chemical remanent magnetization is held by the mineral hematite, another iron oxide. Hematite forms through chemical oxidation reactions of other minerals in the rock including magnetite. Redbeds, clastic sedimentary rocks (such as sandstones) are red because of hematite that formed during sedimentary diagenesis. The CRM signatures in redbeds can be quite useful and they are common targets in magnetostratigraphy studies.

Isothermal Remanent Magnetization

Remanence that is acquired at a fixed temperature is called *isothermal remanent magnetization (IRM)*. Remanence of this sort is not useful for paleomagnetism, but it can be acquired as a result of lightning strikes. Lightning-induced remanent magnetization can be distinguished by its high intensity and rapid variation in direction over scales of centimeters.

IRM is often induced in drill cores by the magnetic field of the steel core barrel. This contaminant is generally parallel to the barrel, and most of it can be removed by heating up to about 400 °C or demagnetizing in a small alternating field. In the laboratory, IRM is induced by applying fields of various strengths and is used for many purposes in rock magnetism.

Viscous Remanent Magnetization

Viscous remanent magnetization, also known as viscous magnetization, is remanence that is acquired by ferromagnetic materials by sitting in a magnetic field for some time. The natural remanent magnetization of an igneous rock can be altered by this process. This is generally an unwanted component and some form of stepwise demagnetization must be used to remove it.

Viscous remanent magnetization is the result of jumps between magnetic states driven by thermal fluctuations. In the simplest case, there is a single characteristic time τ called the *thermal relaxation time*. If the starting magnetization is M_0 and the magnetization in equilibrium is M_{eq}, the magnetization after a time t is:

$$M(t) = M_0 + \left(M_{eq} - M_0\right)\exp(-t/\tau).$$

If M_0 is a natural remanent magnetization (NRM) acquired in the Earth's field at one time, and M_{eq} is the equilibrium magnetization in a different field at some later time, the change in remanence $M(t) - M_0$ is the VRM.

The above equation is applicable to the simplest single domain magnets, those described by the Stoner–Wohlfarth model. The relaxation time τ depends on factors such as the size of the magnet and its magnetic anisotropy. Rocks have collections of magnetic minerals with varying size and anisotropy, and therefore a broad spectrum of relaxation times. The magnetization tends to have a logarithmic dependence on time, so the rate of change is often represented by a *viscosity coefficient*:

$$S = \frac{\partial M}{\partial \log t}.$$

Significance for Paleomagnetism

Paleomagnetists are interested in the *primary* natural remanent magnetization (NRM) in a rock, acquired when the rock was originally formed. Viscous remanent magnetization is regarded as noise. Any component of the NRM that is in the direction of the present Earth's field is suspect because it may have been acquired since the last geomagnetic reversal. VRM is often removed by the first steps in a stepwise thermal demagnetization of the NRM.

VRM may also be acquired in the laboratory while measuring the NRM. To avoid this, paleomagnetists make their measurements in a magnetically shielded environment. Often the magnetometer is housed in a room with walls made of mu-metal.

Paleomagnetic Poles

The inclination and declination of a dipolar magnetic field change with position on the globe. But the position of the magnetic pole of a geocentric dipole is independent of observing locality. For many purposes, comparison of results between various observing localities is facilitated by determining a pole position. This pole position is simply the geographic location of the projection of the negative end of the dipole onto the Earth's surface, as shown in figure.

Calculation of a pole position is a navigational problem in spherical trigonometry that uses the dipole formula to determine the distance traveled from observing locality to pole position. Sign conventions for geographic locations are as follows:

- Latitudes increase from −90° at south geographic pole to 0° at equator and to +90° at the north geographic pole.
- Longitudes east of the Greenwich meridian are positive, while westerly longitudes are negative.

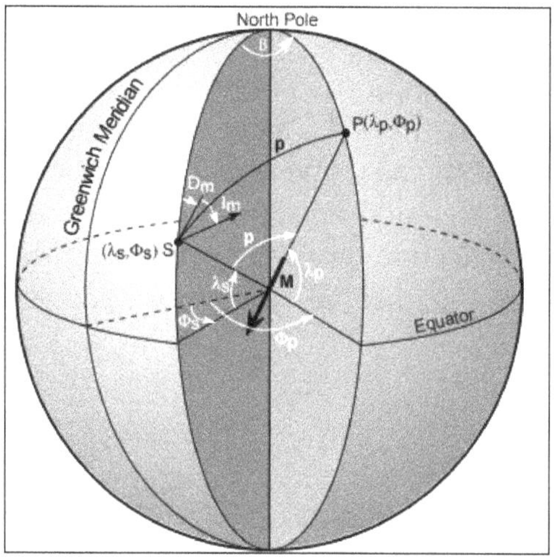

Determination of magnetic pole position from a magnetic field direction. Site location is at S (λ_s, φ_s); site-mean magnetic field direction is I_m, D_m; M is the geocentric dipole that can account for the observed magnetic field direction; P is the magnetic pole at (λ_p, φ_p); p is the magnetic colatitude (angular distance from S to P); North Pole is the north geographic pole; β is the difference in longitude between the magnetic pole and the site.

Figure illustrates how a pole position (λ_p, φ_p) is calculated from a site-mean direction (I_m, D_m) measured at a particular site (λ_s, φ_s). The first step is to determine the magnetic colatitude, p, which is the greatcircle distance from site to pole. From the dipole formula:

$$p = \cot^{-1}\left(\frac{\tan I_m}{2}\right) = \tan^{-1}\left(\frac{2}{\tan I_m}\right)$$

Pole latitude is given by:

$$\lambda_p = \sin^{-1}(\sin\lambda_s \cos p + \cos\lambda_s \sin p \cos D_m)$$

The longitudinal difference between pole and site is denoted by β, is positive toward the east, and is given by:

$$\beta = \sin^{-1}\left(\frac{\sin p \sin D_m}{\cos\lambda_p}\right)$$

At this point in the calculation, there are two possibilities for pole longitude. If:

$$\cos p \geq \sin\lambda_s \sin\lambda_p$$

Then,

$$\phi_p = \phi_s + \beta$$

But if,

$$\cos p < \sin \lambda_s \sin_p$$

Then,

$$\phi_p = \phi_s + 180^\circ - \beta$$

Any site-mean direction I_m, D_m has an associated confidence limit α_{95}. This circular confidence limit about the site-mean direction is transformed (mapped by the dipole formula) into an ellipse of confidence about the calculated pole position. The semi-axis of the ellipse of confidence has an angular length along the site-to-pole great circle given by:

$$dp = \alpha_{95}\left(\frac{1+3\cos^2 p}{2}\right)$$

The semi-axis perpendicular to the great circle is given by:

$$dm = \alpha_{95}\left(\frac{\sin p}{\cos I_m}\right)$$

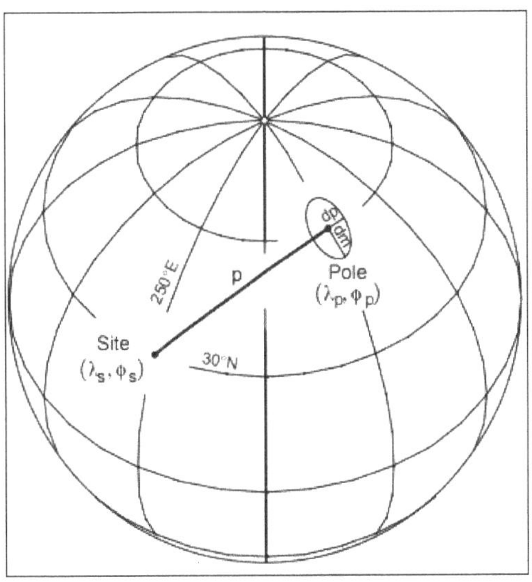

Ellipse of confidence about magnetic pole position. p is the magnetic colatitude; dp is the semi-axis of the confidence ellipse along the greatcircle path from site to pole; dm is the semi-axis of the confidence ellipse perpendicular to that great-circle path. The projection (for this and all global projections to follow) is orthographic with latitude and longitude grid in 30° increments.

As an example calculation, consider a site-mean direction I_m = 45°, D_m = 25° with α_{95} = 5.0° observed at location λ_s = 30°N, φ_s = 250°E (= 110°W). The colatitude, p, given by Equation $p = \cot^{-1}\left(\frac{\tan I_m}{2}\right) = \tan^{-1}\left(\frac{2}{\tan I_m}\right)$ is 63.4°. From Equation $\lambda_p = \sin^{-1}(\sin\lambda_s \cos p + \cos\lambda_s \sin p \cos D_m)$, the pole latitude, λ_p, is 67.8°N, and the angle β from Equation $\beta = \sin^{-1}\left(\frac{\sin p \sin D_m}{\cos\lambda_p}\right)$ is 86.2°. The product sin λ_s sin λ_p = 0.463, while cos p = 0.448, so cos p < sin λ_s sin λ_p, and the pole longitude is given by Equation $\phi_p = \phi_s + 180^\circ - \beta$ as φ_p = 342.7 °E. The pole is illustrated in figure below. Using Equations $dp = \alpha_{95}\left(\frac{1+3\cos^2 p}{2}\right)$ and $dm = \alpha_{95}\left(\frac{\sin p}{\cos I_m}\right)$, the confidence ellipse about the pole has dp = 4.0° and dm = 6.3°.

Types of Poles

The calculation scheme just described yields the position of the north geomagnetic pole, assuming that the observed direction is produced by a geocentric dipole. The geomagnetic field is more complex than a simple geocentric dipole. The present geomagnetic field is composed of a dominant dipolar field and a higher-order nondipole field. In addition, the geomagnetic field changes with time. To deal with these spatial and temporal complications, various types of magnetic poles have been defined. These magnetic poles are determined from different kinds of observations, and the distinctions between them are important.

Geomagnetic Pole

For the present geomagnetic field, it is possible to examine globally distributed observations and determine the best-fitting geocentric dipole. The pole position of that best-fitting dipole is the geomagnetic pole. For the year 1980, the north geomagnetic pole was located at approximately 79°N, 289°E in the Canadian Arctic Islands. For determination of the geomagnetic pole position, globally distributed observations are required to "average out" the nondipole field. An observation of the magnetic field direction at a single location cannot be used because the observed direction would, in general, be affected by the nondipole field. Thus a pole position calculated on the basis of a single observation at a particular location is not expected to coincide with the geomagnetic pole. For example, the present magnetic field direction in Tucson, Arizona ($\lambda_s \approx$ 32°N, $\varphi_s \approx$ 249°E) is I ≈ 60°, D ≈ 14°, and the resulting pole position is $\lambda_p \approx$ 76°N, $\varphi_p \approx$ 297°E, substantially removed from the present geomagnetic pole.

Virtual Geomagnetic Pole

Any pole position that is calculated from a single observation of the direction of the geomagnetic field is called a virtual geomagnetic pole. This is the position of the pole of a geocentric dipole that can account for the observed magnetic field direction at one location and at one point in time. A VGP can be calculated from an observation of the present geomagnetic field direction at a particular locality. If VGPs are determined from many globally distributed observations of the present geomagnetic field, these VGPs are scattered about the present geomagnetic pole. In paleomagnetism, a site-mean ChRM direction is a record of the past geomagnetic field direction at the sampling site location during the ideally short interval of time over which the ChRM was acquired. Thus a pole position calculated from a single site-mean ChRM direction is a virtual geomagnetic pole.

Paleomagnetic Pole

Because of nondipole components, a site-mean VGP is not expected to coincide with the geomagnetic pole at the time the ChRM was acquired. In theory, the geomagnetic pole in ancient times could be determined by paleomagnetic investigation of globally distributed rocks of equivalent age. In practice, dating techniques are sufficiently precise to allow such geomagnetic pole determinations only for the past few thousand years. This direct technique obviously could not be extended to rocks older than about 5 Ma because continental drift has changed the relative positions of observing localities. The only practical solution to averaging out effects of the nondipole field is to time average the field for an interval of time covering the periodicities of secular variation of the nondipole field.

The dipolar geomagnetic field undergoes secular variation, causing the geomagnetic pole to random walk about the rotation axis with periodicities dominantly from 103 to 104 yr. The geocentric axial dipole hypothesis states that, if geomagnetic secular variation has been adequately sampled, the average position of the geomagnetic pole coincides with the rotation axis. Thus a set of paleomagnetic sites magnetized over about 104 to 105 yr should yield an average pole position (average of site-mean VGPs) coinciding with the rotation axis. Pole positions calculated with these criteria satisfied are called paleomagnetic poles. The term paleomagnetic pole implies that the pole position has been determined from a paleomagnetic data set that has averaged geomagnetic secular variation and thus gives the position of the rotation axis with respect to the sampling area at the time the ChRM was acquired.

Procedures for calculating paleomagnetic poles have changed during the past decade. Previously, the approach was to calculate a formation-mean direction by using Fisher statistics to average the site-mean directions from a geological formation. The formation-mean direction then was used to calculate the paleomagnetic pole. A 95% confidence ellipse for the paleomagnetic pole was determined from the α95 circle about the formation-mean direction. This pole position was reported as the paleomagnetic pole from the formation, and the error ellipse was used as an estimate of precision.

The α_{95} circle of confidence about a mean direction is mapped by the dipole formula into an ellipse of confidence about the calculated pole. Similarly, a circular distribution of directions is mapped into an elliptical distribution of VGPs calculated from those directions. But conversely, a circular distribution of VGPs implies that the distribution of directions yielding those VGPs is elliptical. So site-mean directions or site-mean VGPs (but not both) might be circularly distributed about their respective means. Analyses of large paleomagnetic data sets (from rocks up to a few million years in age) indicate that distributions of site-mean VGPs are more nearly circularly distributed about the mean pole position than are sitemean directions about the formation-mean direction. Consequently, most paleomagnetic poles are now determined in the following manner: (1) From each site-mean ChRM direction, a site-mean VGP is calculated. (2) The set of VGPs then is used to find the mean pole position (paleomagnetic pole) by Fisher statistics, treating each VGP as a point on the unit sphere. The procedure for determining the mean pole position is the same as for determining a mean direction except that VGP latitude is substituted for inclination and VGP longitude for declination.

Estimates of (between-site) dispersion of the site-mean VGPs are obtained by using the same procedures applied to directions. But in this case, N = number of site-mean VGPs; R = vector resultant of the N site-mean VGPs; and the confidence limit applies to the calculated mean pole position. An informal convention has developed in which upper-case letters are used for dispersion estimates of VGPs. K is the best-estimate of the precision parameter κ for the observed distribution of site-mean VGPs; S is the angular dispersion of VGPs (estimated angular standard deviation of VGPs) and is usually estimated; A_{95} is the radius of the 95% confidence circle about the calculated mean pole (the true mean pole lies within A_{95} of the calculated mean pole with 95% confidence).

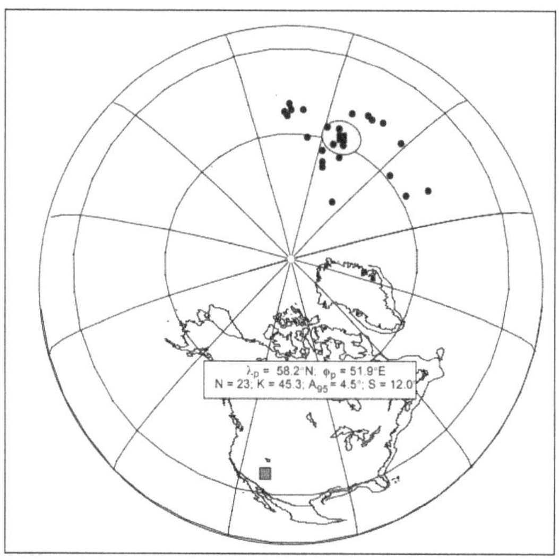

Figure illustrates an example of a paleomagnetic pole (and A_{95} confidence circle) determined from a set of site-mean VGPs. The example is from the Early Jurassic Moenave Formation of northern Arizona and southern Utah. This formation is dominated by red and purple-red sediments, and an example of thermal demagnetization behavior was provided in figure. For most of the 23 sites from which a ChRM was successfully isolated, the site-mean α_{95} was Paleomagnetic pole from the Moenave Formation. Solid circles show the 23 site-mean VGPs averaged to determine the paleomagnetic pole shown by the solid square; the stippled circle about the paleomagnetic pole is the region of 95% confidence with radius A_{95}; the region of sampling is shown by the stippled square; the inset gives the location of the paleomagnetic pole along with statistical parameters.

References

- "Estimated Value of Magnetic Declination". Geomagnetism. NOAA National Geophysical Data Center. Retrieved 6 December 2013
- Mussett, Alan E.; Khan, M. Aftab (2000). Looking into the Earth: An introduction to Geological Geophysics. Cambridge University Press. ISBN 978-0-521-78085-8
- Sources-of-variation-in-the-steady-magnetic-field, geomagnetic-field, science: britannica.com, Retrieved 30 March, 2020
- Sanders, Robert (2003-12-10). "Radioactive potassium may be major heat source in Earth's core". UC Berkeley News. Retrieved 2007-02-28
- Ovchinnikov, I.V. (March 2016). "Introduction to Supersymmetric Theory of Stochastics". Entropy. 18 (4): 108. arXiv:1511.03393. Bibcode:2016Entrp..18..108O. doi:10.3390/e18040108

Plate Tectonics 5

- **Tectonic Plate**
- **Plate Tectonics**
- **Plate Boundaries**
- **Plate Reconstruction**
- **Subduction**
- **Interactions of Tectonics with other Systems**

Plate tectonics is defined as a scientific theory which describes the large scale motion and movements of the plates of the Earth's lithosphere. A number of forces drive the movement of tectonic plates. Three of the major driving forces of tectonic plates are mantle convection currents, ridge push and slab pull. This chapter discusses the subject of plate tectonics in detail.

Tectonic Plate

A tectonic plate (also called lithospheric plate) is a massive, irregularly shaped slab of solid rock, generally composed of both continental and oceanic lithosphere. Plate size can vary greatly, from a few hundred to thousands of kilometers across; the Pacific and Antarctic Plates are among the largest. Plate thickness also varies greatly, ranging from less than 15 km for young oceanic lithosphere to about 200 km or more for ancient continental lithosphere (for example, the interior parts of North and South America).

How do these massive slabs of solid rock float despite their tremendous weight? The answer lies in the composition of the rocks. Continental crust is composed of granitic rocks which are made up of relatively lightweight minerals such as quartz and feldspar. By contrast, oceanic crust is composed of basaltic rocks, which are much denser and heavier. The variations in plate thickness are nature's way of partly compensating for the imbalance in the weight and density of the two types of crust. Because continental rocks are much lighter, the crust under the continents is much thicker (as much as 100 km) whereas the crust under the oceans is generally only about 5 km thick. Like icebergs, only the tips of which are visible above water, continents have deep "roots" to support their elevations.

Most of the boundaries between individual plates cannot be seen, because they are hidden beneath the oceans. Yet oceanic plate boundaries can be mapped accurately from outer space by measurements from GEOSAT satellites. Earthquake and volcanic activity is concentrated near these boundaries. Tectonic plates probably developed very early in the Earth's 4.6-billion-year history, and they have been drifting about on the surface ever since-like slow-moving bumper cars repeatedly clustering together and then separating.

Types of Tectonic Plates

Tectonic Plates are of two types, namely oceanic crust and continental crust that differ in composition. Tectonic plates are not fixed but move above the molten mantle below them.Tectonic plates form either divergent, transform, or convergent boundaries when they come in contact. Such boundaries are highly susceptible to earthquakes and volcanic eruptions. Orogeny also takes place at such boundaries. Tectonic plates are defined as major and minor plates depending on their size. There are a total of seven major tectonic plates which cover nearly 95% of the Earth's surface.

Major Tectonic Plates by Size

Pacific Plate

The Pacific Plate is estimated to be 103,300,000 square kilometers in size. Found underneath the Pacific Ocean, it is the largest of all tectonic plates. Most of the Pacific Plate is made up of oceanic crust, except for areas around New Zealand and parts of California. The nature of the Pacific Plate was notably responsible for forming the islands of Hawaii. The Hawaiian Islands were originally volcanoes that rose above the water over millions of years to form landmasses. These volcanoes were formed at hot spots in the Pacific Plate. This tectonic plate hosts the Ring of Fire, an area on the floor of the Pacific Ocean where volcanic activity and earthquakes are most active.

North American Plate

The North American Plate is the world's second-largest tectonic plate. It consists of both continental crust and oceanic crust. The plate's continental crust is made up of most of North America and Iceland. The North American plate is responsible for the formation of the Mid-Atlantic Ridge, a mountain chain beneath the Atlantic Ocean. A few hot spots underneath the plate are responsible for active seismic activity, the most famous example of which may be the Yellowstone geyser.

Eurasian Plate

The Eurasian Plate has an estimated area of 67,800,000 square kilometers. It is the third-largest of the major tectonic plates. Most of the continents of Europe and Asia are in the Eurasian Plate. Several geological formations can be found on this plate, the most prominent of which is the Himalayan Range. The Himalayan mountains formed as a result of the collision between the Eurasian Plate and the Indian Plate. The Eurasian Plate is a geologically active plate, with volcanoes and earthquakes occurring in its territory.

African Plate

The African plate is the fourth largest tectonic plate with an estimated area of 61,300,000 square kilometers. Most of the African continent is on the African Plate. The African Plate also includes substantial parts of the Indian and Atlantic Oceans. The Plate is slowly splitting at the East African Rift Valley which runs from the Red Sea to Kenya. Notably, the Italian island of Sicily is also a part of the African Plate.

Antarctic Plate

The Antarctic Plate encompasses the entire continent of Antarctica, as well as the nearby oceans. It is the fifth-largest plate on earth. It is also the world's southernmost plate.

Indo-Australian Plate

The Indo-Australian Plate was formed out of a merger of the Australian and Indian plates millions of years ago. When the Eurasian Plate and the Indo-Australian plates collided many many years ago, the Himalayan mountains were formed. Some scientists, however, believe that the Indian Plate and the Australian Plate are separate plates, and have been so for millions of years.

South American Plate

The South American Plate is a major tectonic plate that covers 43 million square kilometers including South America and the surrounding Atlantic Ocean. Tectonic activity at the boundary between the South American Plate and the Nazca Plate is held responsible for the volcanic activity and orogeny in the region.

Minor Tectonic Plates by Size

Somali Plate

The Somali Plate is a minor tectonic plate encompassing the African country of Somalia. Currently, the Somali Plate is moving away from continental Africa at a very small pace which equates to around 20 millimeters per annum. At this pace, Somalia may separate from Africa after millions of years, leading to the formation of a new continent and ocean.

Nazca Plate

The second largest of all minor plates, the Nazca Plate, stretches for 15.6 square km off the western coast of South America, to the south of the much smaller Cocos Plate.

Philippine Sea Plate

The Philippine Sea Plate comprises of over 5 million square km of ocean space adjacent to the Philippines, in the Philippine Sea. The plate also touches upon both Taiwan and Japan in its northern reaches.

Arabian Plate

The Arabian Plate measures 5 million square km, mostly across the Arabian Peninsula. The plate also includes parts of the Levant.

Caribbean Plate

The Caribbean Plate is found in the Caribbean Sea, as well as the island of Hispaniola, and Central America. It lies to the north of South America and south of the islands of Cuba and Jamaica.

Cocos Plate

The Cocos Plate is a minor plate that stretches for 2.9 million square km. It is geographically located off the coast of western Central America. The plate is around 23 million years old, which is relatively young in tectonic plate terms. The formation of the Cocos Plate can be traced to seafloor spreading, which generally occurs at mid-ocean ranges. The shifting of the Cocos Plate underneath the North American Plate (these movements are called subduction) led to several earthquakes in recent times.

Caroline Plate

The Caroline Plate is a minor plate found in South Asia. It is moving at a speed of around 87 mm every year.

Scotia Plate

The Scotia Plate stretches for 1.6 square km just north of the Antarctic Plate. The majority of the plate is deeply submerged beneath the Scotia Sea.

Burma Plate

As its name suggests, the Burma Plate encompasses the country of Burma.

New Hebrides Plate

The New Hebrides Plate is found in the south Pacific Ocean, where it stretches for 1,100,000 square km. It is closest to the country of Vanuatu.

Bonus: Juan de Fuca Plate

The Juan de Fuca Plate is one of the smallest of tectonic plates. At only 205,000 square km, it is technically not a minor plate but a microplate - but it may be one of the world's most notorious ones. The Juan de Fuca Plate is part of the infamous Ring of Fire, a zone responsible for volcanic activity, orogeny, and earthquakes.

Plate Tectonics

Plate tectonics is a theory dealing with the dynamics of Earth's outer shell—the lithosphere—that revolutionized Earth sciences by providing a uniform context for understanding mountain-building processes, volcanoes, and earthquakes as well as the evolution of Earth's surface and reconstructing its past continents and oceans.

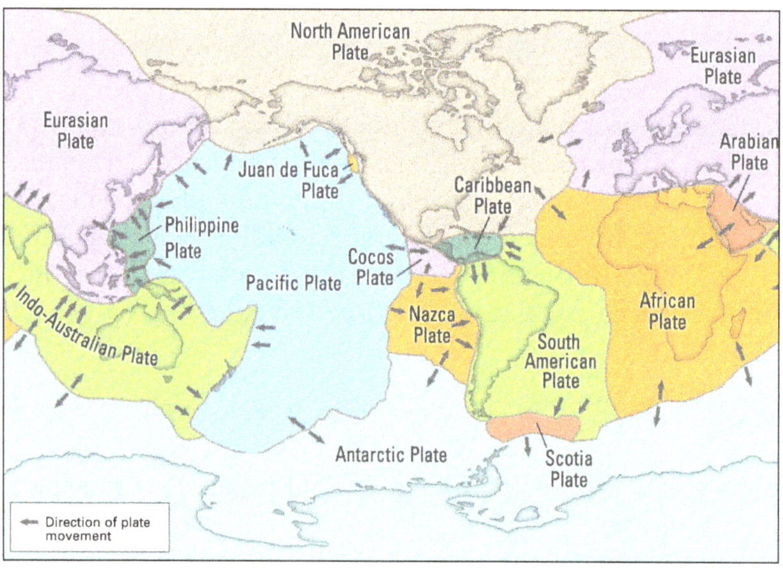

Earth's tectonic plates Map showing Earth's major tectonic plates with arrows depicting the directions of plate movement.

The concept of plate tectonics was formulated in the 1960s. According to the theory, Earth has a rigid outer layer, known as the lithosphere, which is typically about 100 km (60 miles) thick and overlies a plastic (moldable, partially molten) layer called the asthenosphere. The lithosphere is broken up into seven very large continental- and ocean-sized plates, six or seven medium-sized regional plates, and several small ones. These plates move relative to each other, typically at rates of 5 to 10 cm (2 to 4 inches) per year, and interact along their boundaries, where they converge, diverge, or slip past one another. Such interactions are thought to be responsible for most of Earth's seismic and volcanic activity, although earthquakes and volcanoes can occur in plate interiors. Plate motions cause mountains to rise where plates push together, or converge, and continents to fracture and oceans to form where plates pull apart, or diverge. The continents are embedded in the plates and drift passively with them, which over millions of years results in significant changes in Earth's geography.

The theory of plate tectonics is based on a broad synthesis of geologic and geophysical data. It is now almost universally accepted, and its adoption represents a true scientific revolution, analogous in its consequences to quantum mechanics in physics or the discovery of the genetic code in biology. Incorporating the much older idea of continental drift, as well as the concept of seafloor spreading, the theory of plate tectonics has provided an overarching framework in which to describe the past geography of continents and oceans, the processes controlling creation and destruction of landforms, and the evolution of Earth's crust, atmosphere, biosphere, hydrosphere, and climates. During the late 20th and early 21st centuries, it became apparent that plate-tectonic processes profoundly influence the composition of Earth's atmosphere and oceans, serve as a prime cause of long-term climate change, and make significant contributions to the chemical and physical environment in which life evolves.

Principles of Plate Tectonics

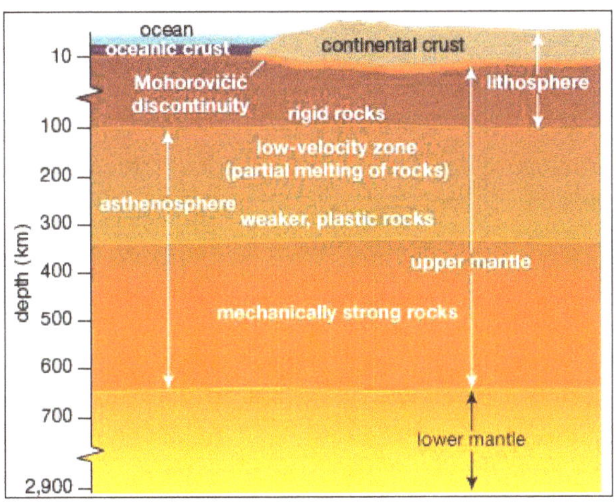

A cross section of Earth's outer layers, from the crust through the lower mantle.

In essence, plate-tectonic theory is elegantly simple. Earth's surface layer, 50 to 100 km (30 to 60 miles) thick, is rigid and is composed of a set of large and small plates. The lithosphere rests on and slides over an underlying partially molten (and thus weaker but generally denser) layer of plastic partially molten rock known as the asthenosphere. Plate movement is possible because the lithosphere-asthenosphere boundary is a zone of detachment. As the lithospheric plates move across Earth's surface, driven by forces as yet not fully understood, they interact along their boundaries, diverging, converging, or slipping past each other. While the interiors of the plates are presumed to remain essentially undeformed, plate boundaries are the sites of many of the principal processes that shape the terrestrial surface, including earthquakes, volcanism, and orogeny (that is, formation of mountain ranges).

The process of plate tectonics may be driven by convection in Earth's mantle, the pull of heavy old pieces of crust into the mantle, or some combination of both. For a deeper discussion of plate-driving mechanisms.

Earth's Layers

Knowledge of Earth's interior is derived primarily from analysis of the seismic waves that propagate through Earth as a result of earthquakes. Depending on the material they travel through, the waves may either speed up, slow down, bend, or even stop if they cannot penetrate the material they encounter.

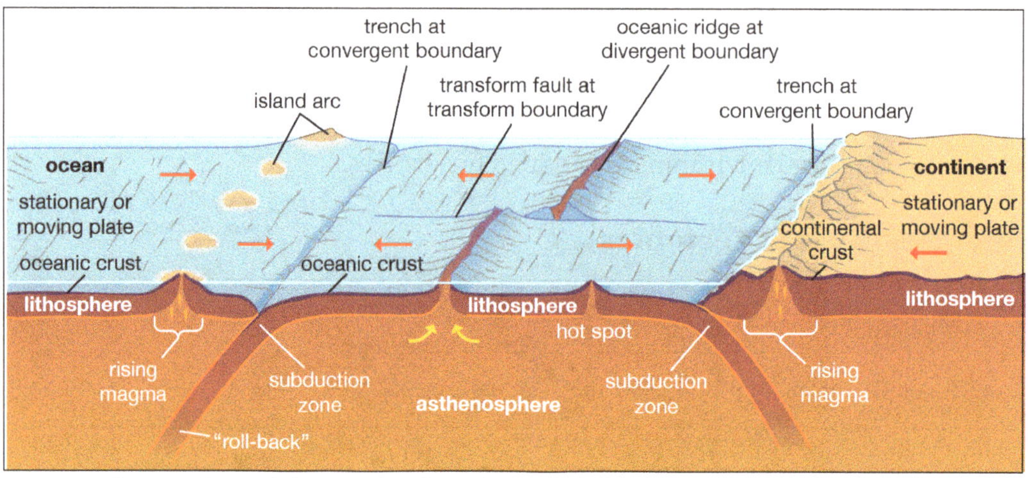

Crustal generation and destruction Three-dimensional diagram showing crustal generation and destruction according to the theory of plate tectonics; included are the three kinds of plate boundaries—divergent, convergent (or collision), and strike-slip (or transform).

Collectively, these studies show that Earth can be internally divided into layers on the basis of either gradual or abrupt variations in chemical and physical properties. Chemically, Earth can be divided into three layers. A relatively thin crust, which typically varies from a few kilometres to 40 km (about 25 miles) in thickness, sits on top of the mantle. (In some places, Earth's crust may be up to 70 km [40 miles] thick.) The mantle is much thicker than the crust; it contains 83 percent of Earth's volume and continues to a depth of 2,900 km (1,800 miles). Beneath the mantle is the core, which extends to the centre of Earth, some 6,370 km (nearly 4,000 miles) below the surface. Geologists maintain that the core is made up primarily of metallic iron accompanied by smaller amounts of nickel, cobalt, and lighter elements, such as carbon and sulfur.

There are two types of crust, continental and oceanic, which differ in their composition and thickness. The distribution of these crustal types broadly coincides with the division into continents and ocean basins, although continental shelves, which are submerged, are underlain by continental crust. The continents have a crust that is broadly granitic in composition and, with a density of about 2.7 grams per cubic cm (0.098 pound per cubic inch), is somewhat lighter than oceanic crust, which is basaltic (i.e., richer in iron and magnesium than granite) in composition and has a density of about 2.9 to 3 grams per cubic cm (0.1 to 0.11 pound per cubic inch). Continental crust is typically 40 km (25 miles) thick, while oceanic crust is much thinner, averaging about 6 km (4 miles) in thickness. These crustal rocks both sit on top of the mantle, which is ultramafic in composition (i.e., very rich in magnesium and iron-bearing silicate minerals). The boundary between the crust (continental or oceanic) and the underlying mantle is known as the Mohorovičić discontinuity (also called Moho), which is named for its discoverer, Croatian seismologist Andrija Mohorovičić. The Moho is clearly defined by seismic studies, which detect an acceleration in seismic waves as they pass from the crust into the denser mantle. The boundary between the mantle and the core is also clearly defined by seismic studies, which suggest that the outer part of the core is a liquid.

The effect of the different densities of lithospheric rock can be seen in the different average elevations of continental and oceanic crust. The less-dense continental crust has greater buoyancy, causing it to float much higher in the mantle. Its average elevation above sea level is 840 metres (2,750 feet), while the average depth of oceanic crust is 3,790 metres (12,400 feet). This density difference creates two principal levels of Earth's surface.

The lithosphere itself includes all the crust as well as the upper part of the mantle (i.e., the region directly beneath the Moho), which is also rigid. However, as temperatures increase with depth, the heat causes mantle rocks to lose their rigidity. This process begins at about 100 km (60 miles) below the surface. This change occurs within the mantle and defines the base of the lithosphere and the top of the asthenosphere. This upper portion of the mantle, which is known as the lithospheric mantle, has an average density of about 3.3 grams per cubic cm (0.12 pound per cubic inch). The asthenosphere, which sits directly below the lithospheric mantle, is thought to be slightly denser at 3.4–4.4 grams per cubic cm (0.12–0.16 pound per cubic inch).

In contrast, the rocks in the asthenosphere are weaker, because they are close to their melting temperatures. As a result, seismic waves slow as they enter the asthenosphere. With increasing depth, however, the greater pressure from the weight of the rocks above causes the mantle to become gradually stronger, and seismic waves increase in velocity, a defining characteristic of the lower mantle. The lower mantle is more or less solid, but the region is also very hot, and thus the rocks can flow very slowly (a process known as creep).

During the late 20th and early 21st centuries, scientific understanding of the deep mantle was greatly enhanced by high-resolution seismological studies combined with numerical modeling and laboratory experiments that mimicked conditions near the core-mantle boundary. Collectively, these studies revealed that the deep mantle is highly heterogeneous and that the layer may play a fundamental role in driving Earth's plates.

At a depth of about 2,900 km (1,800 miles), the lower mantle gives way to Earth's outer core, which is made up of a liquid rich in iron and nickel. At a depth of about 5,100 km (3,200 miles), the outer core transitions to the inner core. Although it has a higher temperature than the outer core, the inner core is solid because of the tremendous pressures that exist near Earth's centre. Earth's inner core is divided into the outer-inner core (OIC) and the inner-inner core (IIC), which differ from one another with respect to the polarity of their iron crystals. The polarity of the iron crystals of the OIC is oriented in a north-south direction, whereas that of the IIC is oriented east-west.

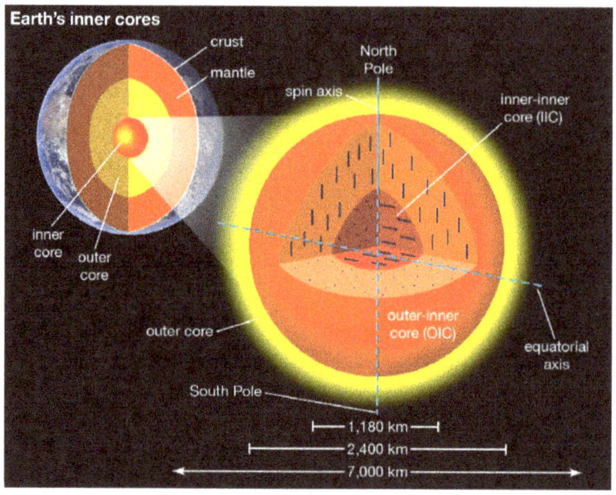

Earth's core The internal layers of Earth's core, including its two inner cores.

Plate Boundaries

Plate boundaries are found at the edge of the lithospheric plates. Wide zones of deformation are usually characteristic of plate boundaries because of the interaction between two plates.

Three types of plate boundaries exist, with a fourth, mixed type, characterized by the way the plates move relative to each other. They are associated with different types of surface phenomena. The different types of plate boundaries are:

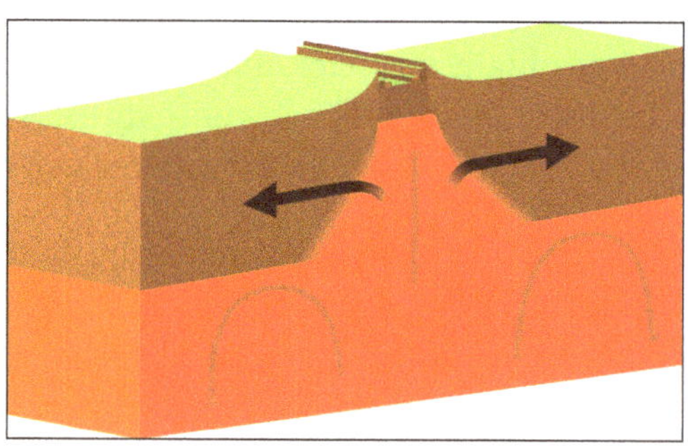

Divergent boundary.

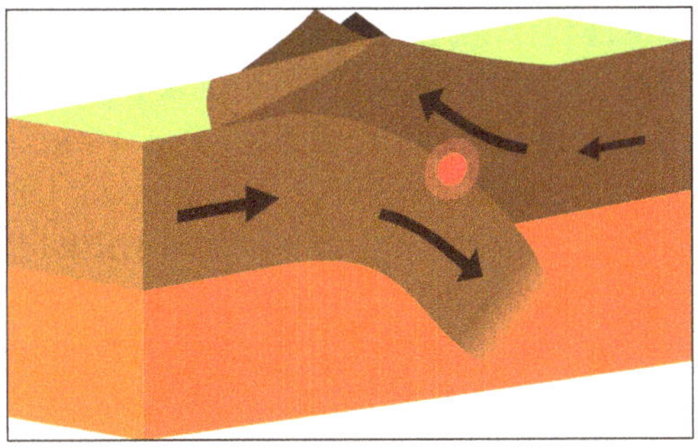

Convergent boundary.

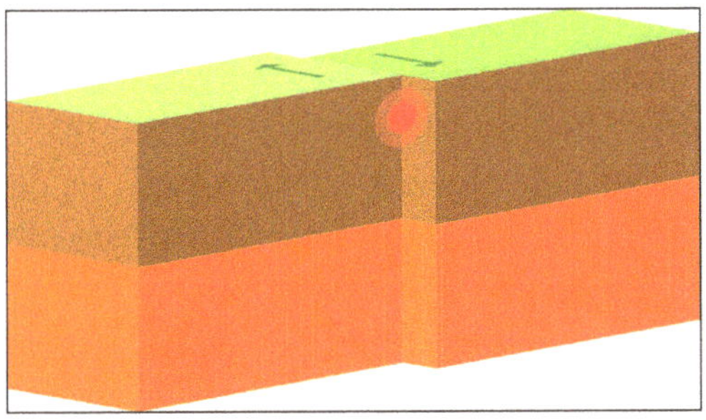

Transform boundary.

- Divergent boundaries (Constructive) occur where two plates slide apart from each other. At zones of ocean-to-ocean rifting, divergent boundaries form by seafloor spreading, allowing for the formation of new ocean basin. As the ocean plate splits, the ridge forms at the spreading center, the ocean basin expands, and finally, the plate area increases causing many small volcanoes and/or shallow earthquakes. At zones of continent-to-continent rifting, divergent boundaries may cause new ocean basin to form as the continent splits, spreads, the central rift collapses, and ocean fills the basin. Active zones of mid-ocean ridges (e.g., the Mid-Atlantic Ridge and East Pacific Rise), and continent-to-continent rifting (such as Africa's East African Rift and Valley and the Red Sea), are examples of divergent boundaries.

- Convergent boundaries (Destructive) (or active margins) occur where two plates slide toward each other to form either a subduction zone (one plate moving underneath the other) or a continental collision. At zones of ocean-to-continent subduction (e.g. the Andes mountain range in South America, and the Cascade Mountains in Western United States), the dense oceanic lithosphere plunges beneath the less dense continent. Earthquakes trace the path of the downward-moving plate as it descends into asthenosphere, a trench forms, and as the subducted plate is heated it releases volatiles, mostly water from hydrous minerals, into the surrounding mantle. The addition of water lowers the melting point of the mantle material above the subducting slab, causing it to melt. The magma that results typically leads to volcanism. At zones of ocean-to-ocean subduction (e.g. Aleutian islands, Mariana Islands, and the Japanese island arc), older, cooler, denser crust slips beneath less dense crust. This motion causes earthquakes and a deep trench to form in an arc shape. The upper mantle of the subducted plate then heats and magma rises to form curving chains of volcanic islands. Deep marine trenches are typically associated with subduction zones, and the basins that develop along the active boundary are often called "foreland basins". Closure of ocean basins can occur at continent-to-continent boundaries (e.g., Himalayas and Alps): collision between masses of granitic continental lithosphere; neither mass is subducted; plate edges are compressed, folded, uplifted.

- Transform boundaries (Conservative) occur where two lithospheric plates slide, or perhaps more accurately, grind past each other along transform faults, where plates are neither created nor destroyed. The relative motion of the two plates is either sinistral (left side toward the observer) or dextral (right side toward the observer). Transform faults occur across a spreading center. Strong earthquakes can occur along a fault. The San Andreas Fault in California is an example of a transform boundary exhibiting dextral motion.

Plate boundary zones occur where the effects of the interactions are unclear, and the boundaries, usually occurring along a broad belt, are not well defined and may show various types of movements in different episodes.

Plate Reconstruction

Plate reconstruction is the process of reconstructing the positions of tectonic plates relative to each other (relative motion) or to other reference frames, such as the earth's magnetic field or groups of hotspots, in the geological past. This helps determine the shape and make-up of ancient supercontinents and provides a basis for paleogeographic reconstructions.

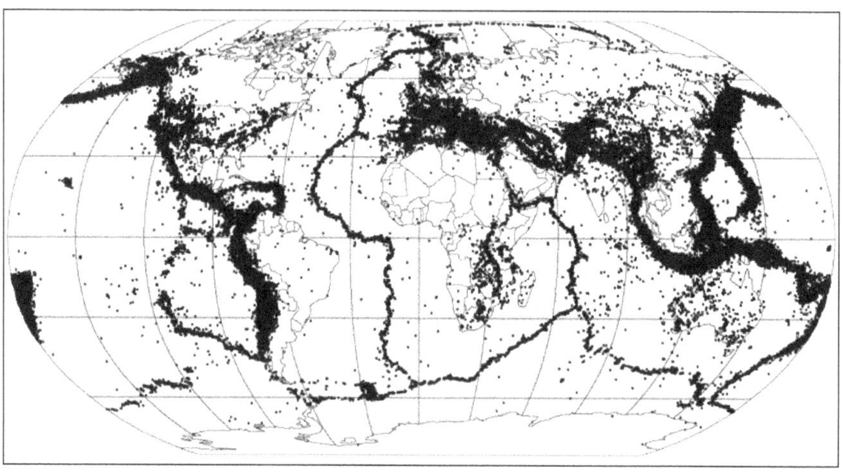

Earthquake epicenters.

An important part of reconstructing past plate configurations is to define the edges of areas of the lithosphere that have acted independently at some time in the past.

Present Plate Boundaries

Most present plate boundaries are easily identifiable from the pattern of recent seismicity. This is now backed up by the use of GPS data, to confirm the presence of significant relative movement between plates.

Past Plate Boundaries

Identifying past (but now inactive) plate boundaries within current plates is generally based on evidence for an ocean that has now closed up. The line where the ocean used to be is normally marked by pieces of the crust from that ocean, included in the collision zone, known as ophiolites. The line across which two plates became joined to form a single larger plate, is known as a suture.

In many orogenic belts, the collision is not just between two plates, but involves the sequential accretion of smaller terranes. Terranes are smaller pieces of continental crust that have been caught up in an orogeny, such as continental fragments or island arcs.

Reference Frames

Plate motions, both those observable now and in the past, are referred ideally to a reference frame that allows other plate motions to be calculated. For example, a central plate, such as the African plate, may have the motions of adjacent plates referred to it. By composition of reconstructions, additional plates can be reconstructed to the central plate. In turn, the reference plate may be reconstructed, together with the other plates, to another reference frame, such as the earth's magnetic field, as determined from paleomagnetic measurements of rocks of known age. A global hotspot reference frame has been postulated but there is now evidence that not all hotspots are necessarily fixed in their locations relative to one another or the earth's spin axis. However, there are groups of such hotspots that appear to be fixed within the constraints of available data, within particular mesoplates.

Euler Poles

The movement of a rigid body, such as a plate, on the surface of a sphere can be described as rotation about a fixed axis (relative to the chosen reference frame). This pole of rotation is known as an Euler pole. The movement of a plate is completely specified in terms of its Euler pole and the angular rate of rotation about the pole. Euler poles defined for current plate motions can be used to reconstruct plates in the recent past (few million years). At earlier stages of earth's history, new Euler poles need to be defined.

Estimating Past Plate Motions

In order to move plates backward in time it is necessary to provide information on either relative or absolute positions of the plates being reconstructed such that an Euler pole can be calculated. These are quantitative methods of reconstruction.

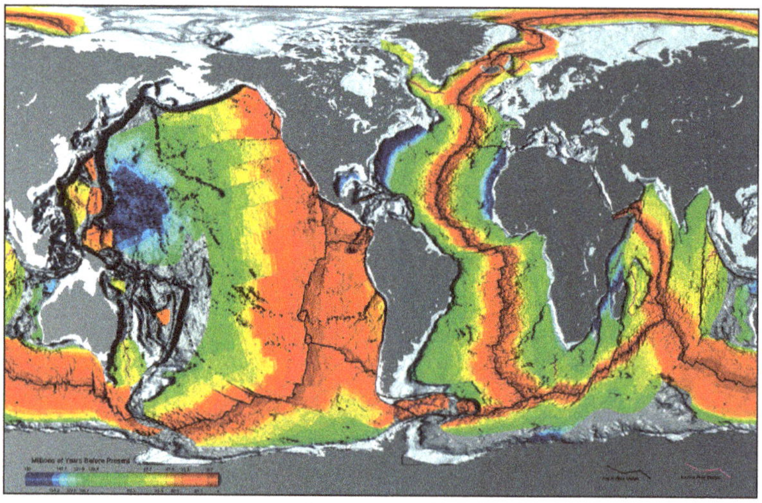

Ages of oceanic lithosphere.

Geometric Matching of Continental Borders

Certain fits between continents, particularly that between South America and Africa, were known long before the development of a theory that could adequately explain them. The reconstruction before Atlantic rifting by Bullard based on a least-squares fitting at the 500 fathom contour still provides the best match to paleomagnetic pole data for the two sides from the middle of Paleozoic to Late Triassic.

Plate Motion from Magnetic Stripes

Plate reconstructions in the recent geological past mainly use the pattern of magnetic stripes in oceanic crust to remove the effects of seafloor spreading. The individual stripes are dated from magnetostratigraphy so that their time of formation is known. Each stripe (and its mirror image) represents a plate boundary at a particular time in the past, allowing the two plates to be repositioned relative to one another. The oldest oceanic crust is Jurassic, providing a lower age limit of about 175 Ma for the use of such data. Reconstructions derived in this way are only relative.

Plate Reconstructions from Paleomagnetism

Paleomagnetic Data: Sampling

Paleomagnetic data are obtained by taking oriented samples of rocks and measuring their remanent magnetizations in the laboratory. Good quality data can be recovered from different rock types. In igneous rocks, magnetic minerals crystallize from the melt, and when the rock is cooled below their Curie temperature, it acquires a thermoremanent magnetization (TRM) in the direction of the Earth's magnetic field. In sedimentary rocks, magnetic grains will align their magnetic moments with the direction of the magnetic field during or soon after the deposition, resulting in a detrital or post-detrital remanent magnetization (DRM). A common difficulty with the use of clastic sediments for defining directions of the magnetic field in the past is that the direction of DRM may rotate toward the bedding plane due to the compaction of sediment, resulting in an inclination, which is shallower than the inclination of the field during the deposition. The inclination flattening error can nevertheless be estimated and corrected for through re-deposition experiments, measurements of magnetic anisotropy, and the use of theoretical models for the dispersion of paleomagnetic directions. Metamorphic rocks are not normally used for paleomagnetic measurements due to the complexities related to the acquisition of remanence, uncertainties in magnetization age, and high magnetic anisotropy.

A typical paleomagnetic study would sample a large number of independent rock units of similar age at nearby locations and collect multiple samples from each unit in order to estimate measurement errors and assess how well the obtained paleomagnetic dataset samples geomagnetic secular variation. Progressive demagnetization techniques are used to identify secondary magnetization components (e.g., magnetic overprints that could have been imparted on the rock due to chemical alteration or reheating) and to isolate the primary magnetization, which records the direction of the magnetic field at the time when the rock was formed. Various rock-magnetic and paleomagnetic tests are normally performed to establish the primary nature of the isolated remanent magnetization. The recovered paleomagnetic directions are used to derive paleomagnetic poles, which provide constrains on the latitudinal position of the crustal block from which the rock samples were taken, and its original orientation with respect to the lines of longitude.

Paleomagnetic Poles

A paleomagnetic pole is defined by taking the average direction of the primary remanent magnetization for the sampled rocks (expressed as the mean declination and inclination) and calculating the position of a geomagnetic pole for the field of a geocentric magnetic dipole that would produce the observed mean direction at the sampled locality in its present geographic coordinates. An alternative way of defining paleomagnetic poles is to calculate a virtual geomagnetic pole (VGP) for each individual rock unit and then estimate the mean location for all VGPs. Fisher statistics on the sphere is normally used to obtain the mean direction of magnetization, or the mean VGP location, and to estimate their uncertainties. Both approaches are used in paleomagnetic studies, but it has been recognized that averaging directions instead of full remanence vectors can lead to biased estimates of the mean direction of the paleomagnetic field, so that the calculation of paleomagnetic poles by averaging VGPs is currently the preferred technique.

Applications to Paleogeographic Reconstructions

Paleogeographic reconstruction of the Pangea supercontinent at the Permo-Triassic Boundary (250 Ma). Top panel: Synthetic APWP for Africa (the south paleomagnetic poles are shown with their 95% uncertainty ovals). The red dot highlights the 250 Ma paleomagnetic pole. APWP data are from Torsvik et al. Middle panel: All continents are assembled in the Pangea configuration at 250 Ma using the estimates of their relative motions, with Africa kept fixed in its present position. The red triangle shows the position of the Euler pole and the red arrow indicates the rotation that would reconstruct the paleomagnetic pole to the south geographic pole. Bottom panel: The Euler rotation has been applied to Pangea, which is now reconstructed paleogeographically. The longitude is arbitrary set to minimize the longitudinal motion of Africa since 250 Ma.

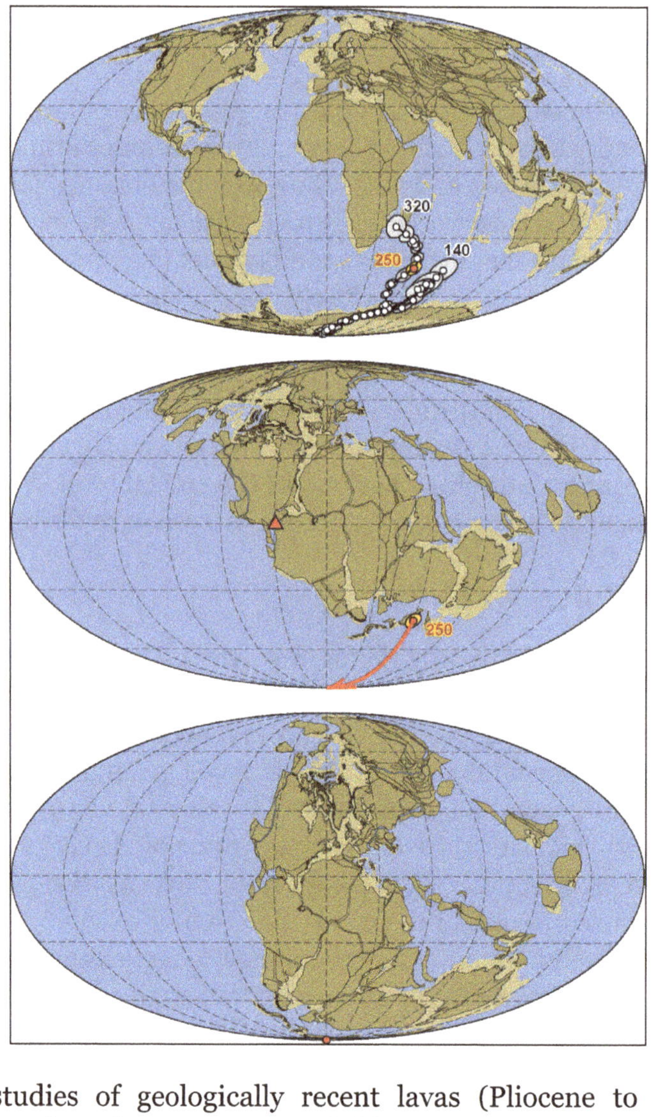

Paleomagnetic studies of geologically recent lavas (Pliocene to Quaternary, 0-5 Ma) indicate that when the geomagnetic field is averaged on time scales of tens of thousands to millions of years – over a time period long enough to fully sample geomagnetic secular variation, the time-averaged field can be accurately approximated by the field of a geocentric axial dipole (GAD) – that is, a magnetic dipole placed in the center of the Earth and aligned with the Earth's rotation axis. Hence, if a paleomagnetic dataset has sampled enough time to average secular variation, the paleomagnetic pole derived from it can be interpreted as an estimate for the location of the geographic pole with respect to the sampling locality fixed in the present geographic position.

The difference between the paleomagnetic pole and the present geographic pole reflects the paleogeographic position of the crustal block containing the sampled area at the time when the studied rocks were formed, including its original latitude (paleolatitude) and orientation. Under the assumption that the mean paleomagnetic direction corresponds to that of the GAD field, the paleolatitude of the sampling location (λ) can be derived from the inclination (I) of the mean direction using a simple equation:

$$\tan(\lambda) = \frac{1}{2} \cdot \tan(I)$$

The mean declination (D) gives the sense and amount of rotation about a vertical axis passing through the sampling area, which needs to be applied to restore its original orientation with respect to the lines of longitude. The paleolatitude for any specific location belonging to the same crustal block can be computed as 90° minus the angular distance between this location and the paleomagnetic pole, and the local vertical axis rotation can be estimated by computing the declination expected from the position of the pole. Thus, a paleomagnetic pole defines the paleo-latitudinal position and orientation of the entire tectonic block at a specific time in the past. However, because the GAD field is azimuthally symmetric about the Earth's rotation axis, the pole doesn't set any constraint on the absolute longitude. From the perspective of paleomagnetic directions, the GAD field has the same values of inclination and declination along a line of constant latitude at all longitudes, so that any conceivable longitude would be an equally viable option for the reconstruction of a tectonic element if its paleogeographic position is constrained by paleomagnetic data alone.

Considering that a paleomagnetic pole approximates the position of the geographic pole with respect to the continent or geologic terrane from which it was determined, the paleolatitude and orientation can be restored by finding a rotation (Euler pole and rotation angle) that reconstructs the paleomagnetic pole to the geographic pole, and applying this rotation to the continent or terrane. By doing so, the crustal block and its paleomagnetic pole are reconstructed using the same Euler rotation, so that they do not move relative to each other, the paleomagnetic pole is placed at the geographic pole, and the crustal block is correctly restored in latitude and orientation (i.e., with respect to the geographic pole). Noting that a further rotation around the geographic pole will only change the longitude of the block, but its latitude and orientation with respect to the lines of longitude will not be affected, the absolute paleolongitude cannot be determined in reconstructions based on paleomagnetism. However, relative longitudes of different crustal blocks can be defined using other types of geological and geophysical data constraining relative motions of tectonic plates, including the histories of seafloor spreading recorded my marine magnetic anomalies, matching of continental borders and geologic terranes, and paleontological data.

Apparent Polar Wander Paths

Poles from different ages in a single continent, lithospheric plate, or any other tectonic block can be used to construct an apparent polar wander path (APWP). If paths from adjacent crustal fragments are identical, this is taken to indicate that there has been no relative movement between them during the period covered by the path. Divergence of APW paths indicates that the areas in question have acted independently in the past with the point of divergence marking the time at which they became joined. Combined or synthetic APWPs can be constructed by rotating paleomagnetic poles from different plates into the reference frame fixed to a single plate, using estimates of relative plate motions. For the times postdating the assembly of Pangea, synthetic APWPs are often constructed in the reference frame fixed to the African plate because Africa has occupied a central position in the Pangea configuration and has been dominantly surrounded by spreading ridges after the Pangea breakup, which commenced in the early Jurassic.

Longitude Constraints

For a single lithospheric plate, the APWP reflects the motion of the plate with respect to the geographic pole (changes in latitude) and changes of its orientation with respect to paleomeridians. The longitudes of paleogeographic reconstructions based on APWPs are uncertain, but it has been argued that the uncertainty can be minimized by selecting a reference plate that is expected to move the least in longitude from the consideration of the plate tectonics theory and by linking the reconstructions of the remaining plates to this reference plate using the estimates of relative plate motion. For example, and it was shown that assuming no significant longitudinal motion of Africa since the time of the Pangea assembly results in a reasonable plate tectonic scenario, in which no large, coherent east-west motions of the continental lithosphere are observed in paleogeographic reconstructions.

APWPs can be interpreted as records of a combined signal from two sources of plate motion: (1) motion of lithospheric plates with respect to the Earth's mantle and (2) motion of the entire solid Earth (mantle and lithosphere) with respect to the Earth's rotation axis. The second component is commonly referred to as true polar wander (TPW) and on geologic time scales results from gradual redistribution of mass heterogeneities due to convective motions in the Earth's mantle. By comparing plate reconstructions based on paleomagnetism with reconstructions in the mantle reference frame defined by hotspots for the last 120 Ma, the TPW motions can be estimated, which allows tying paleogeographic reconstructions to the mantle and hence constraining them in paleolongitude. For the earlier times in the Mesozoic and Paleozoic, TPW estimates can be obtained through the analysis of coherent rotations of the continental lithosphere, which allows linking the reconstructed paleogeography to the large-scale structures in the lower mantle, commonly referred to as Large Low Shear-wave Velocity Provinces (LLSVPs). It has been argued that the LLSVPs have been stable over at least the past 300 Ma, and possibly longer, and that the LLSVP margins have served as generation zones for the mantle plumes responsible for eruptions of Large Igneous Provinces (LIPs) and kimberlites. Correlating the reconstructed locations of LIPs and kimberlites with the margins of LLSVPs using the estimated TPW rotations makes it possible to develop a self-consistent model for plate motions relative to the mantle, true polar wander, and the corresponding changes of paleogeography constrained in longitude for the entire Phanerozoic, although the origin and long-term stability of LLSVPs are the subject of the ongoing scientific debate.

Apparent Polar Wander Paths Geometric Parameterizations

Paleomagnetic Euler poles derived by geometrizing apparent polar wander paths (APWPs) potentially allows constraining paleolongitudes from paleomagnetic data. This method could extend absolute plate motion reconstructions deeply into the geologic history as long as there are reliable APWPs.

Hotspot Tracks

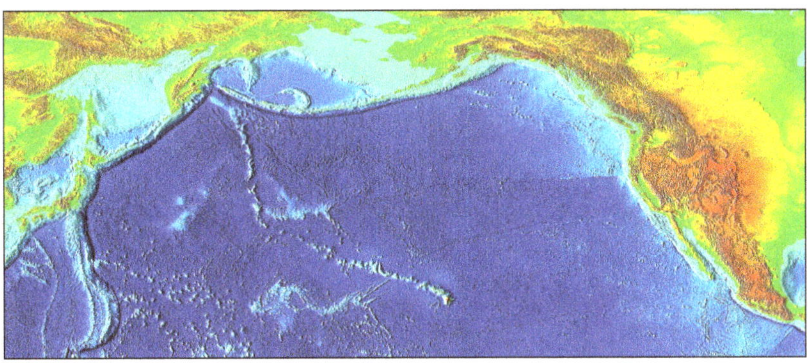

The Hawaiian-Emperor seamount chain.

The presence of chains of volcanic islands and seamounts interpreted to have formed from fixed hotspots allows the plate on which they sit to be progressively restored so that a seamount is moved back over the hotspot at its time of formation. This method can be used back to the Early Cretaceous, the age of the oldest evidence for hotspot activity. This method gives an absolute reconstruction of both latitude and longitude, although before about 90 Ma there is evidence of relative motion between hotspot groups.

Slab Constraints

Once oceanic plates subduct in the lower mantle (slabs), they are assumed to sink in a near-vertical manner. With the help of seismic wave tomography, this can be used to constrain plate reconstructions at first order back to the Permian.

Other Evidence for Past Plate Configurations

Some plate reconstructions are supported by other geological evidence, such as the distribution of sedimentary rock types, the position of orogenic belts and faunal provinces shown by particular fossils. These are semi-quantitative methods of reconstruction.

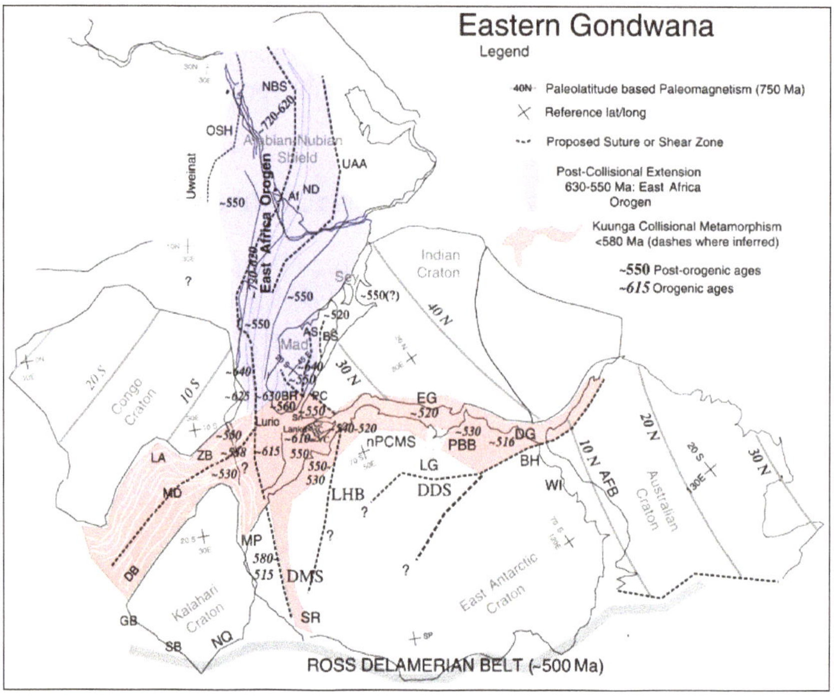

Reconstruction of eastern Gondwana showing position of orogenic belts.

Sedimentary Rock Types

Some types of sedimentary rock are restricted to certain latitudinal belts. Glacial deposits for instance are generally confined to high latitudes, whereas evaporites are generally formed in the tropics.

Faunal Provinces

Oceans between continents provide barriers to plant and animal migration. Areas that have become separated tend to develop their own fauna and flora. This is particularly the case for plants and land animals but is also true for shallow water marine species, such as trilobites and brachiopods, although their planktonic larvae mean that they were able to migrate over smaller deep water areas. As oceans narrow before a collision occurs, the faunas start to become mixed again, providing supporting evidence for the closure and its timing.

Orogenic Belts

When supercontinents break up, older linear geological structures such as orogenic belts may be split between the resulting fragments. When a reconstruction effectively joins up orogenic belts of the same age of formation, this provides further support for the reconstruction's validity.

References

- Major-tectonic-plates-on-earth: worldatlas.com, Retrieved 25 July, 2020
- Glatzmaier, Gary A. (2013). Introduction to Modeling Convection in Planets and Stars: Magnetic Field, Density Stratification, Rotation. Princeton University Press. p. 149. ISBN 978-1-4008-4890-4
- Torsvik, T.H. (2018). "Earth history: A journey in time and space from base to top". Tectonophysics. 760: 297–313. doi:10.1016/j.tecto.2018.09.009. ISSN 0040-1951
- Plate-tectonics, science: britannica.com, Retrieved 11 June, 2020
- "Dumping and Loss overview". Oceans in the Nuclear Age. Archived from the originalon June 5, 2011. Retrieved 18 September 2010
- Continental-reconstructions, science, plate-tectonics-261587: britannica.com, Retrieved 10 July, 2020

PERMISSIONS

All chapters in this book are published with permission under the Creative Commons Attribution Share Alike License or equivalent. Every chapter published in this book has been scrutinized by our experts. Their significance has been extensively debated. The topics covered herein carry significant information for a comprehensive understanding. They may even be implemented as practical applications or may be referred to as a beginning point for further studies.

We would like to thank the editorial team for lending their expertise to make the book truly unique. They have played a crucial role in the development of this book. Without their invaluable contributions this book wouldn't have been possible. They have made vital efforts to compile up to date information on the varied aspects of this subject to make this book a valuable addition to the collection of many professionals and students.

This book was conceptualized with the vision of imparting up-to-date and integrated information in this field. To ensure the same, a matchless editorial board was set up. Every individual on the board went through rigorous rounds of assessment to prove their worth. After which they invested a large part of their time researching and compiling the most relevant data for our readers.

The editorial board has been involved in producing this book since its inception. They have spent rigorous hours researching and exploring the diverse topics which have resulted in the successful publishing of this book. They have passed on their knowledge of decades through this book. To expedite this challenging task, the publisher supported the team at every step. A small team of assistant editors was also appointed to further simplify the editing procedure and attain best results for the readers.

Apart from the editorial board, the designing team has also invested a significant amount of their time in understanding the subject and creating the most relevant covers. They scrutinized every image to scout for the most suitable representation of the subject and create an appropriate cover for the book.

The publishing team has been an ardent support to the editorial, designing and production team. Their endless efforts to recruit the best for this project, has resulted in the accomplishment of this book. They are a veteran in the field of academics and their pool of knowledge is as vast as their experience in printing. Their expertise and guidance has proved useful at every step. Their uncompromising quality standards have made this book an exceptional effort. Their encouragement from time to time has been an inspiration for everyone.

The publisher and the editorial board hope that this book will prove to be a valuable piece of knowledge for students, practitioners and scholars across the globe.

INDEX

A

Absolute Free-fall Gravimeters, 33, 35-37

Atom Interferometer Gravimeter, 33, 36-37

B

Biofilms, 73-78, 80-81, 84-86, 92, 94-95

Biogeochemical Processes, 75, 80, 95-96

Biogeophysics, 42, 73-75, 77, 81, 85-86, 88-89, 92, 94, 96, 98, 100

Borehole Geophysics, 42, 64, 67, 102

Brittle Rocks, 123-124, 129

C

Celestial Bodies, 6, 12, 26

Centrifugal Force, 19, 188

Crustal Rocks, 5, 121-122, 130, 225

D

Data Acquisition Processing, 56, 62

Data Fitting, 55

Data Processing, 43-44, 46, 55-57, 63, 101-102

Data Reduction, 101-102, 200

Detrital Remanent Magnetization, 164, 209-210, 231

Diamond Anvil Cell, 142-145

Ductile Rocks, 126, 129

Dynamic Topography, 118, 146-150

E

Earth Crust, 2, 37, 39-41

Environmental Geophysics, 42-43, 51-52, 55, 63, 112, 117

Environmental Scanning Electron Microscope, 77

Exploration Geophysics, 42, 63-64, 73, 101

F

Falling Corner-cube Gravimeter, 33, 35-37

G

Geodynamic Regimes, 131-135

Geodynamo Theory, 153

Geoelectrical Data, 58-59

Geomagnetic Field, 1, 83, 153-156, 162, 167, 194, 196, 200-202, 214-215, 232

Geometric Geodesy, 6

Geothermal Energy, 2, 39, 140

Geothermal Gradient, 118, 137-142

Gravimetry, 1, 27-31, 41, 102, 107

Gravitational Force, 7, 9-10, 16, 20-23, 188

Gravity Anomalies, 32, 142, 200

Gravity Measurements, 17-18, 20, 34, 59, 102-103

Gravity-microgravity Method, 50

H

Horizontal Plane, 4, 122, 171-172

Horizontal Strata, 126-127

Hydrocarbon Exploration, 28, 65

Hydrogeophysical Inversion, 111, 116

Hydrological Science, 113

I

Induced Polarization, 43, 48, 64-65, 76, 81-82, 98, 102, 107, 114

Interplanetary Magnetic Field, 160, 181

L

Laser Interferometer, 25-26, 33, 36

Lithospheric Yield Stress, 134, 136

Lower Mantle, 2, 223, 225-226, 235

M

Magnetic Dip, 157, 160, 172, 180-181

Magnetic Fields, 3, 21, 164-165, 167, 183-184, 186, 189, 191, 195-196

Magnetic Inclination, 151, 172, 180, 207

Magnetic Signals, 83

Magnetic Susceptibility, 4-5, 43, 51-53, 60, 63, 65, 76, 200

Magnetosphere, 157, 160-162, 167, 170, 201-202

Magnetostratigraphy, 157, 164, 206, 208, 210, 230

Magnetotactic Bacteria, 77-78, 80, 83

Mantle Heating, 134-135
Metallic Iron Minerals, 96, 99-100
Microbial Cells, 74, 76-77, 80-85
Mineral Exploration, 49, 64-65, 91, 101
Multi Anvil Press, 145

N
Near-surface Geophysics, 42, 101, 103
Nonmetallic Minerals, 99-100

O
Oceanic Crust, 101, 139, 146, 148, 206, 218-219, 225, 230
Orogeny, 219-220, 222, 224, 229

P
Paleomagnetic Poles, 151, 211, 215-216, 231-232, 234
Petrophysical Properties, 74, 80
Physical Geodesy, 6
Plate Model, 147-148
Plate Reconstruction, 218, 228

R
Redox Processes, 79, 89-90
Remote Sensing, 43, 64-65, 67, 101-102
Residual Topography, 147-149
Reverse Faults, 124, 130

S
Saturated Zone, 108, 114
Secondary Minerals, 95-97
Seismic Methods, 43-46, 65, 102
Seismic Reflection Technique, 45
Seismic Velocities, 45, 144
Spectral Induced Polarization, 43, 76, 81, 98, 114
Spectroscopy, 76, 81-82, 85, 110, 133, 144-145
Spherical Harmonics, 14-16, 19-20, 148, 168-170
Spontaneous Potential, 76, 102, 114
Spring Gravimeter, 34-36
Spring-based Relative Gravimeters, 34
Stagnant Lid, 132, 134-136
Subduction, 150, 196, 218, 221, 227-228
Surface Waves, 44-45, 47

T
Tectonic Plate, 137, 218-221
Tectonophysics, 42, 104-105, 117
Terrestrial Exoplanets, 118, 130, 133-134
Thermoremanent Magnetization, 164, 201, 208, 231
Topographic, 43, 50, 146, 175

U
Uniform Stress, 119
Unsaturated Zone, 109-110, 114
Ureolytic Bacteria, 95-96, 99-100

V
Vadose Zone, 109-110, 112, 116
Vertical Gradient, 49-50, 61, 89, 200
Viscoelastic-plastic Models, 134-135

www.ingramcontent.com/pod-product-compliance
Lightning Source LLC
LaVergne TN
LVHW080631210926
848485LV00003B/246